STUDENT SOLUTIONS MAN

TO ACCOMPANY

CHARLES P. McKEAGUE

ELEMENTARY

ALGEBRA

FOURTH EDITION

PREPARED BY

PATRICIA K. BEZONA

VALDOSTA STATE COLLEGE

Saunders College Publishing
Harcourt Brace College Publishers
Fort Worth   Philadelphia   San Diego
New York   Orlando   Austin   San Antonio
Toronto   Montreal   London   Sydney   Tokyo

PREFACE

This Student Solutions Manual accompanies *Elementary Algebra*, Fourth Edition, by Charles P. McKeague.

I believe this manual will help students who need more guidance with the mathematics problems than is usually available through class time or individual assistance.

The manual contains solutions to every other odd-numbered problem in the problem sets and chapter reviews, and to all problems in the chapter tests. The solutions should be followed, step-by-step, to solve any problem that is confusing to you. The solutions can be a guide on homework problems.

I would like to thank Charles P. McKeague both for giving me this opportunity and for all his guidance; and Amy Barnett, my editor at Harcourt Brace Jovanovich, for her patience and understanding. My mother, Theo Reed, and grandmother, Elsie Gabardi, spent many hours taking care of my daughter Tammy and were a constant source of encouragement. A special thanks to my husband, Ron, for his support during this project.

PATRICIA K. BEZONA

CONTENTS

Solutions to Selected Problems

Chapter 1

Section 1.1

1. The sum of 7 and 8 is 15.

5. The difference of 8 and 3 is not equal to 6.

9. The sum of x and 1 is equal to 5.

13. $5y < 30$

17. $3y \leq y + 6$

21. $16 < 17$ True, 16 is less than 17.

25. $3 + 2 < 5$
    $5 < 5$ False, 5 is not less than 5 but equal to 5.

29. $3^2 = 3 \cdot 3$ Base 3, exponent 2
    $= 9$

33. $2^3 = 2 \cdot 2 \cdot 2$ Base 2, exponent 3
    $= 8$

37. $2^4 = 2 \cdot 2 \cdot 2 \cdot 2$ Base 2, exponent 4
    $= 16$

41. $11^2 = 11 \cdot 11$ Base 11, exponent 2
    $= 121$

45. $2(3 + 5) = 16$ Twice the sum of 3 and 5 is 16

49. $(5 + 2) \cdot 6 = 7 \cdot 6$ Inside parentheses first
    $= 42$ Multiply

53. $5(4 + 2) = 5(6)$ 5 times the sum of 4 and 2
    $= 30$

57. $(8 + 2)(5 - 3) = 10(2)$ Inside parentheses first
    $= 20$ Multiply

61. $5 + 2(3 \cdot 4 - 1) + 8 = 5 + 2(12 - 1) + 8$ Inside parentheses first
    $= 5 + 2(11) + 8$ Subtract
    $= 5 + 22 + 8$ Multiply
    $= 35$ Add left to right

Section 1.1 continued

65.
$$4 + 8 \div 4 - 2 = 4 + 2 - 2 \qquad \text{Divide}$$
$$= 6 - 2 \qquad \text{Add}$$
$$= 4 \qquad \text{Subtract}$$

69.
$$3 \cdot 8 + 10 \div 2 + 4 \cdot 2 = 24 + 5 + 8 \qquad \text{Multiply and divide left to right}$$
$$= 37 \qquad \text{Add}$$

73.
$$5^2 - 3^2 = 25 - 9 \qquad \text{Simplify numbers with exponents}$$
$$= 16 \qquad \text{Subtract}$$

77.
$$4^2 + 5^2 = 16 + 25 \qquad \text{Simplify numbers with exponents}$$
$$= 41 \qquad \text{Add}$$

81.
$$2 \cdot 10^3 + 3 \cdot 10^2 + 4 \cdot 10 + 5$$
$$= 2 \cdot 1000 + 3 \cdot 100 + 4 \cdot 10 + 5 \qquad \text{Simplify numbers with exponents}$$
$$= 2000 + 300 + 40 + 5 \qquad \text{Multiply}$$
$$= 2345 \qquad \text{Add}$$

85.
$$4[7 + 3(2 \cdot 9 - 8)] = 4[7 + 3(18 - 8)] \qquad \text{Inside parentheses first}$$
$$= 4[7 + 3(10)] \qquad \text{Subtract}$$
$$= 4[7 + 30] \qquad \text{Multiply}$$
$$= 4[37] \qquad \text{Add}$$
$$= 148$$

89.
$$3(4 \cdot 5 - 12) + 6(7 \cdot 6 - 40)$$
$$= 3(20 - 12) + 6(42 - 40) \qquad \text{Inside parentheses first}$$
$$= 3(8) + 6(2) \qquad \text{Subtract}$$
$$= 24 + 12 \qquad \text{Multiply}$$
$$= 36$$

93.
$$5^2 + 3^4 \div 9^2 + 6^2$$
$$= 25 + 81 \div 81 + 36 \qquad \text{Simplify numbers with exponents}$$
$$= 25 + 1 + 36 \qquad \text{Divide}$$
$$= 62$$

97.
$$3(50) - 14 \qquad \text{Tripled means three times, and to loses means to subtract.}$$

Section 1.2

Answers to the odd numbers 1-8 (See page A-1 in the textbook.)

9. $\dfrac{3}{4} = \dfrac{3 \cdot 6}{4 \cdot 6} = \dfrac{18}{24}$

13. $\dfrac{5}{8} = \dfrac{5 \cdot 3}{8 \cdot 3} = \dfrac{15}{24}$

17. $\dfrac{11}{30} = \dfrac{11 \cdot 2}{30 \cdot 2} = \dfrac{22}{60}$

19. -10, $\frac{1}{10}$, $|10| = 10$

23. $-\frac{11}{2}$, $\frac{2}{11}$, $\left|\frac{11}{2}\right| = \frac{11}{2}$

27. $\frac{2}{5}$, $-\frac{5}{2}$, $\left|-\frac{2}{5}\right| = \frac{2}{5}$

31. $-5 < -3$ Less than

35. $|-4| > -|-4|$ Greater than

39. $-\frac{3}{4} < -\frac{1}{4}$ Less than

43. $|8 - 2| = |6|$ Simplify numbers inside symbol
$\qquad\quad = 6$ Absolute value

47. $|7 - 2| - |4 - 2| = |5| - |2|$ Simplify numbers inside symbol
$\qquad\qquad\qquad\quad = 5 - 2$ Absolute value
$\qquad\qquad\qquad\quad = 3$ Subtract

51. $15 - |8 - 2(3 \cdot 4 - 9)| - 10 = 15 - |8 - 2(12 - 9)| - 10$
$\qquad\qquad\qquad\qquad\qquad\quad = 15 - |8 - 2(3)| - 10$ ⎫ Simplify
$\qquad\qquad\qquad\qquad\qquad\quad = 15 - |8 - 6| - 10$ ⎬ innermost
$\qquad\qquad\qquad\qquad\qquad\quad = 15 - |2| - 10$ ⎭ symbols first
$\qquad\qquad\qquad\qquad\qquad\quad = 15 - 2 - 10$ Absolute value
$\qquad\qquad\qquad\qquad\qquad\quad = 3$ Subtract

55. $-10(0) = 1$
$\qquad\quad 0 = 1$ False

59. $-(-15) = 15$
$\qquad 15 = 15$ True

63. $-|-9| = -9$
$\qquad -(9) = -9$
$\qquad\quad -9 = -9$ True

67. $\frac{2}{3} \cdot \frac{4}{5} = \frac{8}{15}$

71. $\frac{1}{4}(5) = \frac{1}{4} \cdot \frac{5}{1}$

$\qquad = \frac{5}{4}$

75. $6(\frac{1}{6}) = \frac{6}{1} \cdot \frac{1}{6}$

$\qquad = \frac{6}{6}$

$\qquad = 1$

79. $(\frac{3}{4})^2 = \frac{3}{4} \cdot \frac{3}{4}$

$\qquad = \frac{9}{16}$

83. $(\frac{1}{10})^4 = \frac{1}{10} \cdot \frac{1}{10} \cdot \frac{1}{10} \cdot \frac{1}{10}$

$\qquad = \frac{1}{10,000}$

Section 1.3

1. $\qquad 3 + 5 = 8$
$\qquad 3 + (-5) = -2$
$\qquad -3 + 5 = 2$
$\qquad -3 + (-5) = -8$

5. $6 + (-3) = 3$

9. $18 + (-32) = -14$

13. $-30 + 5 = -25$

17. $-9 + (-10) = -19$

21. $5 + (-6) + (-7) = -1 + (-7)$ Applying the rule for order of operations
$\qquad\qquad\qquad\qquad = -8$ Add left to right

25. $5 + [6 + (-2)] + (-3) = 5 + (4) + (-3)$ Add inside the brackets first,
$\qquad\qquad\qquad\qquad\qquad = 9 + (-3)$ then left to right.
$\qquad\qquad\qquad\qquad\qquad = 6$

29. $20 + (-6) + [3 + (-9)] = 20 + (-6) + (-6)$ Add inside the brackets first,
$\qquad\qquad\qquad\qquad\qquad\quad = 14 + (-6)$ then left to right
$\qquad\qquad\qquad\qquad\qquad\quad = 8$

Section 1.3 continued

33. $(-9 + 2) + [5 + (-8)] + (-4) = (-9 + 2) + -3 + (-4)$ Add inside the
$ = -7 + (-3) + (-4)$ brackets, then
$ = -10 + (-4)$ the parentheses,
$ = -14$ then left to right.

37. $(-6 + 9) + (-5) + (-4 + 3) + 7 = 3 + (-5) + (-1) + 7$
$ = -2 + (-1) + 7$
$ = -3 + 7$
$ = 4$

41. $9 + 3(-8 + 10) = 9 + 3(2)$ Add inside the parentheses,
$ = 9 + 6$ then left to right.
$ = 15$

45. $2(-4 + 7) + 3(-6 + 8) = 2(3) + 3(2)$ Add inside the parentheses
$ = 6 + 6$ Multiply left to right
$ = 12$ Add

49. $|-3 + 2| + |-4 + 1| = |-1| + |-3|$ Simplify each absolute value
$ = 1 + 3$ Absolute value
$ = 4$ Add

53. $-3 + (-4) = -7$ $-a + (-b) = -(a + b)$

57. $-8 + (-2) = -10$ $-a + (-b) = -(a + b)$

61. $-10 + (-15) = -25$ $-a + (-b) = -(a + b)$

65. $5 + 9 = 14$

69. $[-2 + (-3)] + 10 = -5 + 10$
$ = 5$

73. $? + (-6) = -9$ The answer is -3.
$-3 + (-6) = -9$

77. $10 + (-6) + (-8) = 4 + (-8)$
$ = -4$

Section 1.4

1. $5 - 8 = 5 + (-8)$ Adding the opposite
$ = -3$

5. $5 - 5 = 5 + (-5)$ Adding the opposite
$ = 0$

9. $-4 - 12 = -4 + (-12)$ Adding the opposite
$ = -16$

5

13. $-8 - (-1) = -8 + 1$ Adding the opposite
$\qquad\qquad\quad = -7$

17. $-4 - (-4) = -4 + 4$ Adding the opposite
$\qquad\qquad\quad = 0$

21. $9 - 2 - 3 = 7 - 3$ Subtract left to right
$\qquad\qquad\quad = 4$

25. $-22 + 4 - 10 = -18 - 10$ Add
$\qquad\qquad\qquad\quad = -18 + (-10)$ Adding the opposite
$\qquad\qquad\qquad\quad = -28$

29. $8 - (2 - 3) - 5 = 8 - (-1) - 5$ Adding the opposite
$\qquad\qquad\qquad\qquad = 8 + 1 - 5$ Adding the opposite
$\qquad\qquad\qquad\qquad = 9 - 5$
$\qquad\qquad\qquad\qquad = 4$

33. $5 - (-8 - 6) - 2 = 5 - [-8 + (-6)] - 2$ Adding the opposite
$\qquad\qquad\qquad\qquad\quad = 5 - (-14) - 2$ Adding the opposite
$\qquad\qquad\qquad\qquad\quad = 5 + 14 - 2$
$\qquad\qquad\qquad\qquad\quad = 19 - 2$
$\qquad\qquad\qquad\qquad\quad = 17$

37. $-(3 - 10) - (6 - 3) = -[3 + (-10)] - (6 + -3)$ Adding the opposite
$\qquad\qquad\qquad\qquad\qquad = -(-7) - 3$ Add
$\qquad\qquad\qquad\qquad\qquad = 7 + (-3)$ Adding the opposite
$\qquad\qquad\qquad\qquad\qquad = 4$

41. $5 - [(2 - 3) - 4] = 5 - [2 + (-3) - 4]$ Adding the opposite
$\qquad\qquad\qquad\qquad\quad = 5 - [-1 - 4]$ Adding the opposite
$\qquad\qquad\qquad\qquad\quad = 5 - [-1 + (-4)]$ Adding the opposite
$\qquad\qquad\qquad\qquad\quad = 5 - [-5]$ Adding the opposite
$\qquad\qquad\qquad\qquad\quad = 5 + 5$ Adding the opposite
$\qquad\qquad\qquad\qquad\quad = 10$

45. $2 \cdot 8 - 3 \cdot 5 = 16 - 15$ Multiply
$\qquad\qquad\qquad\quad = 16 + (-15)$ Adding the opposite
$\qquad\qquad\qquad\quad = 1$

49. $5 \cdot 9 - 2 \cdot 3 - 6 \cdot 2 = 45 - 6 - 12$ Multiply
$\qquad\qquad\qquad\qquad\qquad = 45 + (-6) + (-12)$ Adding the opposite
$\qquad\qquad\qquad\qquad\qquad = 39 + (-12)$ Add
$\qquad\qquad\qquad\qquad\qquad = 27$

53. $2 \cdot 3^2 - 5 \cdot 2^2 = 2 \cdot 9 - 5 \cdot 4$ Simplify exponents
$\qquad\qquad\qquad\qquad = 18 - 20$ Multiply
$\qquad\qquad\qquad\qquad = 18 + (-20)$ Adding the opposite
$\qquad\qquad\qquad\qquad = -2$

57. $|4 - 8| - |2 - 3| = |4 + (-8)| - |2 + (-3)|$ Simplify in the absolute value symbols

$\qquad\qquad\qquad = |-4| - |-1|$ Add

$\qquad\qquad\qquad = 4 - 1$

$\qquad\qquad\qquad = 3$

61. $|-3 - 4| - |-2 - 5| = |-3 + (-4)| - |-2 + 5|$ Adding the opposite

$\qquad\qquad\qquad = |-7| - |-7|$ Add

$\qquad\qquad\qquad = 7 - 7$ Absolute value

$\qquad\qquad\qquad = 0$

65. $12 - (-8) = 12 + 8$ Adding the opposite

$\qquad\qquad = 20$ Add

69. $[4 + (-5)] - 17 = -1 - 17$ Add

$\qquad\qquad = -1 + (-17)$ Adding the opposite

$\qquad\qquad = -18$

73. $-8 - 5 = -8 + (-5)$ Adding the opposite

$\qquad\qquad = -13$

77. $8 - ? = -2$ Answer is 10

$8 - 10 = -2$ Remember, $8 + (-10) = -2$

81. $\$1,500$ In savings account

$\$1,500 - \$730$ Withdrawal

Section 1.5

1. $3 + 2 = 2 + 3$ Commutative property of addition

5. $4 + x = x + 4$ Commutative property of addition

9. $(3 + 1) + 2 = (1 + 3) + 2$ Commutative and associative properties of addition

$\qquad\qquad = 1 + (3 + 2)$

13. $3(x + 2) = 3(2 + x)$ Commutative property of addition

17. $4(xy) = 4(yx)$ Commutative property of multiplication

21. $3(x + 2) = 3(x) + 3(2)$

$\qquad\qquad = 3x + 6$

25. $3(0) = 0$

29. $10(1) = 10$

33. $(x + 2) + 7 = x + (2 + 7)$

$\qquad\qquad = x + 9$

41. $\frac{1}{3}(3x) = [\frac{1}{3}(3)]x$

$= 1x$

$= x$

45. $\frac{3}{4}(\frac{4}{3}x) = (\frac{3}{4} \cdot \frac{4}{3})x$

$= 1x$

$= x$

49. $8(x + 2) = 8(x) + 8(2)$

$= 8x + 16$

53. $4(y + 1) = 4(y) + 4(1)$

$= 4y + 4$

57. $2(3a + 7) = 2(3a) + 2(7)$

$= 6a + 14$

61. $\frac{1}{2}(3x - 6) = \frac{1}{2}(3x) - \frac{1}{2}(6)$ Remember, $-\frac{1}{2}(6) = -\frac{1}{2}(\frac{6}{1}) = -\frac{6}{2} = -3$

$= \frac{3}{2}x - 3$

65. $3(x + y) = 3x + 3y$

69. $6(2x + 3y) = 6(2x) + 6(3y)$

$= 12x + 18y$

73. $\frac{1}{2}(6x + 4y) = \frac{1}{2}(6x) + \frac{1}{2}(4y)$

$= 3x + 2y$

77. $2(3x + 5) + 2 = 2(3x) + 2(5) + 2$

$= 6x + 10 + 2$

$= 6x + 12$

81. No, because a man would not put on his shoes and then his socks.

85. Answers will vary, examples $16 \div 2 \neq 2 \div 16$, $8 \div 4 \neq 4 \div 8$.

Section 1.6

1. $7(-6) = -42$ Different signs give a negative answer.

5. $-8(2) = -16$ Different signs give a negative answer.

8

9. $-11(-11) = 121$ Like signs give a positive answer.

13. $-3(-4)(-5) = 12 \, (-5)$ Multiply
               $= -60$ Multiply

17. $(-7)^2 = (-7)(-7)$ Definition of exponents
         $= 49$ Multiply

21. $-2(2 - 5) = -2(-3)$ Subtract
             $= 6$ Multiply

25. $(4 - 7)(6 - 9) = (-3)(-3)$ Subtract
                   $= 9$ Multiply

29. $-3(-6) + 4(-1) = 18 + (-4)$ Multiply
                   $= 14$ Add

33. $4(-3)^2 + 5(-6)^2 = 4(9) + 5(36)$ Definition of exponents
                $= 36 + 180$ Multiply
                $= 216$ Add

37. $6 - 4(8 - 2) = 6 - 4(6)$ Simplify inside parentheses
             $= 6 - 24$ Multiply
             $= -18$ Subtract

41. $4(2 + 3) - 5(4 + 7) = 4(5) - 5(11)$ Simplify inside parentheses
                  $= 20 - 55$ Multiply
                  $= -35$ Subtract

45. $3|-3 + 10| - 4|-3 + 7| = 3|7| - 4|4|$ Inside absolute value first
                      $= 3(7) - 4(4)$ Absolute value
                      $= 21 - 16$ Multiply
                      $= 5$ Subtract

49. $7 - 2[-6 - 4(-3)] = 7 - 2[-6 + 12]$ Multiply
                $= 7 - 2(6)$ Add
                $= 7 - 12$ Multiply
                $= -5$ Subtract

53. $8 - 6[-2(-3 - 1) + 4(-2 - 3)] = 8 - 6[-2(-4) + 4(-5)]$ Subtract
                             $= 8 - 6[8 + (-20)]$ Multiply
                             $= 8 - 6(-12)$ Add
                             $= 8 + 72$ Multiply
                             $= 80$ Add

57. $-8\left(\dfrac{1}{2}\right) = -\dfrac{8}{1}\left(\dfrac{1}{2}\right)$

               $= -\dfrac{8}{2}$

               $= -4$

61. $-3(-\frac{1}{3}) = -\frac{3}{1}(-\frac{1}{3})$

$\qquad = \frac{3}{3}$

$\qquad = 1$

65. $(-\frac{2}{3})^3 = (-\frac{2}{3})(-\frac{2}{3})(-\frac{2}{3})$

$\qquad = \frac{4}{9}(-\frac{2}{3})$

$\qquad = -\frac{8}{27}$

69. $-7(-6x) = [-7(-6)]x \qquad$ Associative property
$\qquad\quad = 42x \qquad\qquad\qquad$ Multiply

73. $-4(-\frac{1}{4}x) = [-4(-\frac{1}{4})]x \qquad$ Associative property

$\qquad\qquad = 1x \qquad\qquad\qquad$ Multiply
$\qquad\qquad = x$

77. $-\frac{1}{2}(3x - 6) = -\frac{1}{2}(3x) -\frac{1}{2}(-6) \qquad$ Distributive property

$\qquad\qquad = -\frac{3}{2}x + \frac{6}{2} \qquad\qquad$ Multiply

$\qquad\qquad = -\frac{3}{2}x + 3$

81. $-5(3x + 4) - 10 = -5(3x) + (-5)(4) - 10 \qquad$ Distributive property
$\qquad\qquad\qquad = -15x - 20 - 10 \qquad\qquad$ Multiply
$\qquad\qquad\qquad = -15x - 30 \qquad\qquad\qquad$ Subtract

85. $2(-4x) = [2(-4)]x \qquad$ Associative property
$\qquad\quad = -8x \qquad\qquad\qquad$ Multiply

Section 1.7

1. $\frac{8}{-4} = -2 \qquad$ Unlike signs give a negative answer

5. $\frac{-7}{21} = -\frac{1}{3} \qquad$ Reduce by dividing the numerator and denominator by 7

9. $\frac{-6}{-42} = \frac{1}{7} \qquad$ Reduce by dividing the numerator and denominator by 6

Section 1.7 continued

13. $-3 + 12 = 9$

17. $-3(12) = -36$

21. $\dfrac{4}{5} \div \dfrac{3}{4} = \dfrac{4}{5} \cdot \dfrac{4}{3}$ Rewrite as multiplication by the reciprocal

       $= \dfrac{16}{15}$ Rewrite

25. $\dfrac{10}{13} \div \left(-\dfrac{5}{4}\right) = \dfrac{10}{13}\left(-\dfrac{4}{5}\right)$ Rewrite as multiplication by the reciprocal

       $= \dfrac{40}{65}$ Multiply

       $= -\dfrac{8}{13}$ Reduce by dividing numerator and denominator by 5

29. $-\dfrac{3}{4} \div \left(-\dfrac{3}{4}\right) = -\dfrac{3}{4}\left(-\dfrac{4}{3}\right)$ Rewrite as multiplication by the reciprocal

       $= \dfrac{12}{12}$ Multiply

       $= 1$ Divide

33. $\dfrac{-5(-5)}{-15} = \dfrac{25}{-15}$ Multiply

       $= -\dfrac{5}{3}$ Divide numerator and denominator by 5

37. $\dfrac{27}{4 - 13} = \dfrac{27}{-9}$ Simplify denominator

       $= -\dfrac{3}{1}$ Divide numerator and denominator by 9

       $= -3$ Divide

41. $\dfrac{-3 + 9}{2(5) - 10} = \dfrac{6}{10 - 10}$ Simplify numerator and denominator separately

       $= \dfrac{6}{0}$ Undefined

Section 1.7 continued

45. $\dfrac{27 - 2(-4)}{-3(5)} = \dfrac{27 + 8}{-15}$ Simplify numerator and denominator separately

$= -\dfrac{35}{15}$ Add numerators

$= -\dfrac{7}{3}$ Divide numerator and denominator by 5

49. $\dfrac{5^2 - 2^2}{-5 + 2} = \dfrac{25 - 4}{-5 + 2}$ Definition of exponents

$= \dfrac{21}{-3}$ Simplify numerator and denominator separately

$= -7$ Divide

53. $\dfrac{(5 + 3)^2}{-5^2 - 3^2} = \dfrac{8^2}{-25 - 9}$ Remember $-5^2 = -(5)^2 = -(5 \cdot 5) = -25$

$= \dfrac{64}{-34}$

$= -\dfrac{32}{17}$ Divide numerator and denominator by 2

57. $\dfrac{-4 \cdot 3^2 - 5 \cdot 2^2}{-8(7)} = \dfrac{-4 \cdot 9 - 5 \cdot 4}{-56}$ Definition of exponents

$= \dfrac{-36 - 20}{-56}$ Multiply

$= \dfrac{-56}{-56}$ Subtract

$= 1$ Divide numerator and denominator by -56

61. $\dfrac{7 - [(2 - 3) - 4]}{-1 - 2 - 3} = \dfrac{7 - [(-1) - 4]}{-6}$ Simplify numerator and denominator separately

$= \dfrac{7 - [-5]}{-6}$ Subtract

$= \dfrac{7 + 5}{-6}$ Subtract

$= \dfrac{12}{-6}$ Add

$= -2$ Divide numerator and denominator by -2

Section 1.7 continued

65. $\dfrac{3(-5 - 3) + 4(7 - 9)}{5(-2) + 3(-4)} = \dfrac{3(-8) + 4(-2)}{-10 + (-12)}$ Simplify numerator and denominator separately

$= \dfrac{-24 + (-8)}{-22}$ Multiply

$= \dfrac{-32}{-22}$ Add

$= \dfrac{16}{11}$ Divide numerator and denominator by -2

69. $\dfrac{|2 - 9| - |5 - 7|}{10 - 15} = \dfrac{|-7| - |-2|}{-5}$ Simplify numerator and denominator separately

$= \dfrac{7 - 2}{-5}$ Absolute value

$= \dfrac{5}{-5}$

$= -1$ Divide the numerator and denominator by -5

73. $\dfrac{5|-3 + 7| - 6|-4 + 2|}{-1 - 5} = \dfrac{5|4| - 6|-2|}{-6}$ Simplify numerator and denominator separately

$= \dfrac{5(4) - 6(2)}{-6}$ Absolute value

$= \dfrac{20 - 12}{-6}$

$= \dfrac{8}{-6}$

$= -\dfrac{4}{3}$ Divide the numerator and denominator by 2

77. $\dfrac{?}{-5} = 2$ Answer is -10

$\dfrac{-10}{-5} = 2$

81. $\dfrac{-20}{4} - 3 = -5 - 3$

$= -8$

Section 1.8

1. 0, 1

13

5. $-3, -2.5, 0, 1, \dfrac{3}{2}, \sqrt{15}$

9. π

13. False.   (A number is either rational <u>or</u> irrational.)

17. True

21. Prime

25. $144 = 12 \cdot 12$
$= 4 \cdot 3 \cdot 4 \cdot 3$
$= 2 \cdot 2 \cdot 3 \cdot 2 \cdot 2 \cdot 3$
$= 2^4 \cdot 3^2$

29. $105 = 3 \cdot 35$
$= 3 \cdot 5 \cdot 7$

33. $385 = 5 \cdot 77$
$= 5 \cdot 7 \cdot 11$

37. $420 = 10 \cdot 42$
$= 2 \cdot 5 \cdot 6 \cdot 7$
$= 2 \cdot 5 \cdot 2 \cdot 3 \cdot 7$
$= 2^2 \cdot 3 \cdot 5 \cdot 7$

41. $\dfrac{105}{165} = \dfrac{3 \cdot 5 \cdot 7}{3 \cdot 5 \cdot 11}$

$= \dfrac{\cancel{3} \cdot \cancel{5} \cdot 7}{\cancel{3} \cdot \cancel{5} \cdot 11}$

$= \dfrac{7}{11}$

45. $\dfrac{385}{455} = \dfrac{5 \cdot 7 \cdot 11}{5 \cdot 7 \cdot 13}$

$= \dfrac{\cancel{5} \cdot \cancel{7} \cdot 11}{\cancel{5} \cdot \cancel{7} \cdot 13}$

$= \dfrac{11}{13}$

49. $\dfrac{205}{369} = \dfrac{5 \cdot 41}{9 \cdot 41}$

$= \dfrac{5 \cdot \cancel{41}}{9 \cdot \cancel{41}}$

$= \dfrac{5}{9}$

Section 1.8 continued

53. $6^3 = (2 \cdot 3)^3$
$\quad\quad = 2^3 \cdot 3^3$

57. $3 \cdot 8 + 3 \cdot 7 + 3 \cdot 5 = 24 + 21 + 15$
$\quad\quad\quad\quad\quad\quad\quad\quad\quad\quad = 60$
$\quad\quad\quad\quad\quad\quad\quad\quad\quad\quad = 2^2 \cdot 3 \cdot 5$

Section 1.9

1. $\dfrac{3}{6} + \dfrac{1}{6} = \dfrac{3+1}{6}$ Add numerators; keep the same denominator.

$\quad\quad\quad = \dfrac{4}{6}$ The sum of 3 and 1 is 4.

$\quad\quad\quad = \dfrac{2}{3}$ Reduce to lowest terms.

5. $-\dfrac{1}{4} + \dfrac{3}{4} = \dfrac{-1+3}{4}$ Add numerator; keep the same denominator

$\quad\quad\quad = \dfrac{2}{4}$ The sum of -1 and 3 is 2.

$\quad\quad\quad = \dfrac{1}{2}$ Reduce to lowest terms

9. $\dfrac{1}{4} + \dfrac{2}{4} + \dfrac{3}{4} = \dfrac{1+2+3}{4}$

$\quad\quad\quad\quad = \dfrac{6}{4}$

$\quad\quad\quad\quad = \dfrac{3}{2}$

13. $\dfrac{1}{10} - \dfrac{3}{10} - \dfrac{4}{10} = \dfrac{1-3-4}{10}$

$\quad\quad\quad\quad\quad = \dfrac{-6}{10}$

$\quad\quad\quad\quad\quad = -\dfrac{3}{5}$ Unlike signs give a negative answer.

17. $\dfrac{1}{8} + \dfrac{3}{4} = \dfrac{1}{8} + \dfrac{3 \cdot 2}{4 \cdot 2}$ LCD = 8

$\quad\quad\quad = \dfrac{1}{8} + \dfrac{6}{8}$

$\quad\quad\quad = \dfrac{7}{8}$

21. $\dfrac{4}{9} + \dfrac{1}{3} = \dfrac{4}{9} + \dfrac{1 \cdot 3}{3 \cdot 3}$ LCD = 9

$\phantom{\dfrac{4}{9} + \dfrac{1}{3}} = \dfrac{4}{9} + \dfrac{3}{9}$

$\phantom{\dfrac{4}{9} + \dfrac{1}{3}} = \dfrac{7}{9}$

25. $-\dfrac{3}{4} + 1 = -\dfrac{3}{4} + \dfrac{1}{1}$

$\phantom{-\dfrac{3}{4} + 1} = -\dfrac{3}{4} + \dfrac{1 \cdot 4}{1 \cdot 4}$ LCD = 4

$\phantom{-\dfrac{3}{4} + 1} = -\dfrac{3}{4} + \dfrac{4}{4}$

$\phantom{-\dfrac{3}{4} + 1} = \dfrac{-3 + 4}{4}$

$\phantom{-\dfrac{3}{4} + 1} = \dfrac{1}{4}$

29. $\dfrac{x}{4} + \dfrac{1}{5} = \dfrac{x}{4}\dfrac{5}{5} + \dfrac{1 \cdot 4}{5 \cdot 4}$ LCD = 20

$\phantom{\dfrac{x}{4} + \dfrac{1}{5}} = \dfrac{5x}{20} + \dfrac{4}{20}$

$\phantom{\dfrac{x}{4} + \dfrac{1}{5}} = \dfrac{5x + 4}{20}$

33. $\dfrac{5}{12} - \left(-\dfrac{3}{8}\right) = \dfrac{5}{12} + \dfrac{3}{8}$ LCD = 24

$\phantom{\dfrac{5}{12} - \left(-\dfrac{3}{8}\right)} = \dfrac{5 \cdot 2}{12 \cdot 2} + \dfrac{3 \cdot 3}{8 \cdot 3}$

$\phantom{\dfrac{5}{12} - \left(-\dfrac{3}{8}\right)} = \dfrac{10}{24} + \dfrac{9}{24}$

$\phantom{\dfrac{5}{12} - \left(-\dfrac{3}{8}\right)} = \dfrac{29}{24}$

37. $\dfrac{a}{10} + \dfrac{1}{100} = \dfrac{a \cdot 10}{10 \cdot 10} + \dfrac{1}{100}$ LCD = 100

$\phantom{\dfrac{a}{10} + \dfrac{1}{100}} = \dfrac{10a}{100} + \dfrac{1}{100}$

$\phantom{\dfrac{a}{10} + \dfrac{1}{100}} = \dfrac{10a + 1}{100}$

41. $\dfrac{17}{30} + \dfrac{11}{42} = \dfrac{17 \cdot 7}{30 \cdot 7} + \dfrac{11 \cdot 5}{42 \cdot 5}$ LCD = 210

$\qquad\qquad = \dfrac{119}{210} + \dfrac{55}{210}$

$\qquad\qquad = \dfrac{174}{210}$

$\qquad\qquad = \dfrac{29}{35}$

45. $\dfrac{13}{126} - \dfrac{13}{180} = \dfrac{13 \cdot 10}{126 \cdot 10} - \dfrac{13 \cdot 7}{180 \cdot 7}$ LCD = 1,260

$\qquad\qquad = \dfrac{130}{1260} - \dfrac{91}{1260}$

$\qquad\qquad = \dfrac{39}{1260}$

$\qquad\qquad = \dfrac{13}{420}$

49. $\dfrac{4}{y} + \dfrac{2}{3} + \dfrac{1}{2} = \dfrac{4 \cdot 6}{y \cdot 6} + \dfrac{2 \cdot 2y}{3 \cdot 2y} + \dfrac{1 \cdot 3y}{2 \cdot 3y}$ LCD = 6y

$\qquad\qquad = \dfrac{24}{6y} + \dfrac{4y}{6y} + \dfrac{3y}{6y}$

$\qquad\qquad = \dfrac{24 + 4y + 3y}{6y}$

$\qquad\qquad = \dfrac{7y + 24}{6y}$

53. $\dfrac{3}{7} + 2 + \dfrac{1}{9} = \dfrac{3}{7} + \dfrac{2}{1} + \dfrac{1}{9}$ LCD = 63

$\qquad\qquad = \dfrac{3 \cdot 9}{7 \cdot 9} + \dfrac{2 \cdot 63}{1 \cdot 63} + \dfrac{1 \cdot 7}{9 \cdot 7}$

$\qquad\qquad = \dfrac{27}{63} + \dfrac{126}{63} + \dfrac{7}{63}$

$\qquad\qquad = \dfrac{27 + 126 + 7}{63}$

$\qquad\qquad = \dfrac{160}{63}$

Chapter 1 Review

1. The difference of 4 and 9 is not equal to 8.

5. $-7 + (-10) = -17$

9. $9 - (-3) = 9 + 3$
$= 12$

13. $2(-8 \cdot 3x) = 2(-24x)$
$= -48x$

17. See page A-3 in the textbook.

21. See page A-3 in the textbook.

25. $\left| -\dfrac{4}{5} \right| = \dfrac{4}{5}$

29. opposite is -6, reciprocal is $\dfrac{1}{6}$

33. $\left(\dfrac{2}{5} \right)\left(\dfrac{3}{7} \right) = \dfrac{2 \cdot 3}{5 \cdot 7}$

$= \dfrac{6}{35}$

37. $\left(-\dfrac{4}{5} \right)\left(\dfrac{25}{16} \right) = -\dfrac{4 \cdot 25}{5 \cdot 16}$

$= -\dfrac{100}{80}$

$= -\dfrac{5}{4}$

41. $-18 + (-20) = -38$

45. $(-5) + (-10) + (-7) = -15 + (-7)$
$= -22$

49. $(-21) + 40 + (-23) + 5 = 19 + (-23) + 5$
$= -4 + 5$
$= 1$

53. $14 - (-8) = 14 + 8$
$= 22$

57. $4 - 9 - 15 = 4 + (-9) + (-15)$
$= -5 + (-15)$
$= -20$

Chapter 1 Review continued

61. $5 - (-10 - 2) - 3 = 5 - [-10 + (-2)] - 3$
$= 5 - (-12) - 3$
$= 5 + 12 + (-3)$
$= 17 + (-3)$
$= 14$

65. $20 - [-(10 - 3) - 8] - 7$
$= 20 - [-(7) - 8] - 7$
$= 20 - [-7 + (-8)] - 7$
$= 20 - (-15) - 7$
$= 20 + 15 + (-7)$
$= 35 + (-7)$
$= 28$

69. $4(-3) = -12$

73. $(-1)(-3)(-1)(-4)$
$= 3(-1)(-4)$
$= (-3)(-4)$
$= 12$

77. $\dfrac{-9}{36} = -\dfrac{9}{36}$

$= -\dfrac{1}{4}$

81. $4 \cdot 5 + 3 = 20 + 3$
$= 23$

85. $2^3 - 4 \cdot 3^2 + 5^2$
$= 8 - 4 \cdot 9 + 25$
$= 8 - 36 + 25$
$= 8 + (-36) + 25$
$= -28 + 25$
$= -3$

89. $20 + 8 \div 4 + 2 \cdot 5$
$= 20 + 2 + 10$
$= 22 + 10$
$= 32$

93. $-4(-5) + 10 = 20 + 10 = 30$

97. $3(4 - 7)^2 - 5(3 - 8)^2$
$= 3[4 + (-7)]^2 - 5[3 + (-8)]^2$
$= 3(-3)^2 - 5(-5)^2$
$= 3(9) - 5(25)$
$= 27 - 125$
$= 27 + (-125)$
$= -98$

101. $\dfrac{4(-3)}{-6} = \dfrac{-12}{-6}$

$\qquad\qquad = \dfrac{12}{6}$

$\qquad\qquad = 2$

105. $\dfrac{15 - 10}{6 - 6} = \dfrac{5}{0}$ $\qquad$ undefined

109. $\dfrac{2(-7) + (-11)(-4)}{7 - (-3)} = \dfrac{-14 + 44}{10}$

$\qquad\qquad\qquad\qquad = \dfrac{30}{10}$

$\qquad\qquad\qquad\qquad = 3$

113. $(4 + y) + 2 = (y + 4) + 2$ $\qquad$ Commutative property of addition

117. $(4 + 2) + y = (4 + y) + 2$ $\qquad$ Commutative and associative properties of addition

121. $\dfrac{1}{9}(9x) = (\dfrac{1}{9} \cdot 9)x$

$\qquad\qquad = (\dfrac{1}{9} \cdot \dfrac{9}{1})x$

$\qquad\qquad = \dfrac{9}{9}x$

$\qquad\qquad = 1x$

$\qquad\qquad = x$

125. $\dfrac{1}{2}(5x - 6) = \dfrac{1}{2}(5x) - \dfrac{1}{2}(6)$

$\qquad\qquad\qquad = \dfrac{1}{2}(\dfrac{5x}{1}) - \dfrac{1}{2}(\dfrac{6}{1})$

$\qquad\qquad\qquad = \dfrac{5x}{2} - \dfrac{6}{3}$

$\qquad\qquad\qquad = \dfrac{5x}{2} - 2$

129. $\sqrt{7}$, π

133. $840 = 84 \cdot 10$

$\qquad\quad = 7 \cdot 12 \cdot 2 \cdot 5$

$\qquad\quad = 7 \cdot 3 \cdot 4 \cdot 2 \cdot 5$

$\qquad\quad = 7 \cdot 3 \cdot 2 \cdot 2 \cdot 2 \cdot 5$

$\qquad\quad = 2^3 \cdot 3 \cdot 5 \cdot 7$

137. $\dfrac{x}{3} - \dfrac{1}{2} = \dfrac{x \cdot 2}{3 \cdot 2} - \dfrac{1 \cdot 3}{2 \cdot 3}$ LCD = 6

$\qquad\qquad = \dfrac{2x}{6} - \dfrac{3}{6}$

$\qquad\qquad = \dfrac{2x - 3}{6}$

Chapter 1 Test

1. The sum of 3 and 4 equals 7.

2. The difference of 8 and 2 is less than 10.

3. $5^2 + 3(9 - 7) + 3^2 = 25 + 3(9 - 7) + 9$ Simplify number with exponent
   $= 25 + 3(2) + 9$ Simplify parentheses
   $= 25 + 6 + 9$ Multiply
   $= 40$ Add

4. $10 - 6 \div 3 + 2^3 = 10 - 6 \div 3 + 8$ Simplify number with exponent
   $= 10 - 2 + 8$ Divide
   $= 8 + 8$ Add or subtract left to right
   $= 16$ Add

5. $4, -\frac{1}{4}, |-4| = 4$

6. $-\frac{3}{4}, \frac{4}{3}, \left|\frac{3}{4}\right| = \frac{3}{4}$

7. $3 + (-7) = -4$

8. $|-9 + (-6)| + |-3 + 5| = |-15| + |2|$
   $= 15 + 2$
   $= 17$

9. $-4 - 8 = -4 + (-8)$
   $= -12$

10. $9 - (7 - 2) - 4 = 9 - (5) - 4$
    $= 4 - 4$
    $= 0$

11. c. Associative property of Addition. This problem changes the <u>grouping</u> of the numbers with addition.

12. e. Distributive property. Multiplication of 3 is <u>distributed</u> over (x + 5).

13. d. Associative property of Multiplication. The problem changes the <u>grouping</u> of the numbers with multiplication.

14. a. Commutative property of Addition. This problem changes the <u>order</u> of (x + 5) and 7.

15. $-3(7) = -21$ Unlike signs give a negative answer.

16. $-4(8)(-2) = -32(-2)$ Unlike signs give a negative answer.
          $= 64$ Like signs give a positive answer.

17. $8(-\frac{1}{4}) = \frac{8}{1}(-\frac{1}{4})$ Unlike signs give a negative answer

          $= -\frac{8}{4}$ Multiply

          $= -2$ Reduce to lowest terms

18. $(-\frac{2}{3})^3 = (-\frac{2}{3})(-\frac{2}{3})(-\frac{2}{3})$

          $= \frac{4}{9}(-\frac{2}{3})$ Like signs give a positive answer

          $= -\frac{8}{27}$ Unlike signs give a negative answer

19. $-3(-4) - 8 = 12 - 8$ Multiply
          $= 4$ Subtract

20. $5(-6)^2 - 3(-2)^3 = 5(36) - 3(-8)$ Simplify number with exponents
          $= 180 + 24$ Multiply
          $= 204$ Subtract

21. $7 - 3(2 - 8) = 7 - 3(-6)$ Simplify within parentheses
          $= 7 + 18$ Multiply
          $= 25$

22. $4 - 2[-3(-1 + 5) + 4(-3)] = 4 - 2[-3(4) + 4(-3)]$ Simplify innermost
                                                              symbols
                              $= 4 - 2[-12 + -12]$ Multiply in brackets
                              $= 4 - 2[-24]$ Add in brackets
                              $= 4 + 48$ Multiply
                              $= 52$

23. $\frac{4(-5) - 2(7)}{-10 - 7} = \frac{-20 - 14}{-10 - 7}$ Multiply

          $= \frac{-34}{-17}$ Add

          $= 2$ Reduce to lowest terms

24. $\frac{2(-3 - 1) + 4(-5 + 2)}{-3(2) - 4} = \frac{2(-4) + 4(-3)}{-3(2) - 4}$ Simplify parentheses

          $= \frac{-8 + -12}{-6 - 4}$ Multiply

          $= \frac{-20}{-10}$ Add

          $= 2$ Reduce to lowest terms

25. $3 + (5 + 2x) = (3 + 5) + 2x$
$= 8 + 2x$

26. $-2(-5x) = (-2 \cdot -5)x$
$= 10x$

27. $2(3x + 5) = 2(3x) + 2(5)$
$= 6x + 10$

28. $-\frac{1}{2}(4x - 2) = -\frac{1}{2}(4x) + -\frac{1}{2}(-2)$

$= -\frac{4}{2}x + 1$

$= -2x + 1$

29. $1, -8$

30. $1, 1.5, \frac{3}{4}, -8$

31. $\sqrt{2}$

32. All of them.

33. $592 = 2 \cdot 296$
$= 2 \cdot 2 \cdot 148$
$= 2 \cdot 2 \cdot 2 \cdot 74$
$= 2 \cdot 2 \cdot 2 \cdot 2 \cdot 37$
$= 2^4 \cdot 37$

34. $1,340 = 2 \cdot 670$
$= 2 \cdot 2 \cdot 335$
$= 2 \cdot 2 \cdot 5 \quad 67$
$= 2^2 \cdot 5 \cdot 67$

35. $\frac{5}{15} + \frac{11}{42}$ LCD $= 2 \cdot 3 \cdot 5 \cdot 7 = 210$

$= \frac{5}{15} \cdot \frac{14}{14} + \frac{11}{42} \cdot \frac{5}{5}$ $15 = 3 \cdot 5$
$42 = 2 \cdot 3 \cdot 7$

$= \frac{70}{210} + \frac{55}{210}$

$= \frac{125}{210}$

$= \frac{25}{4}$ Divide   numerator and denominator by 5

36. $\dfrac{5}{x} + \dfrac{3}{2}$

$\qquad = \dfrac{5}{x} \cdot \dfrac{2}{2} + \dfrac{3}{2} \cdot \dfrac{x}{x} \qquad$ LCD = 2x

$\qquad = \dfrac{10}{2x} + \dfrac{3x}{2x}$

$\qquad = \dfrac{3x + 10}{2x}$

37. $8 + -3 = 5$

38. $-24 - 2 = -26$

39. $(-5)(-4) = 20$

40. $\dfrac{-24}{-2} = 12$

Chapter 2

Section 2.1

1. $3x - 6x = (3 - 6)x$ Distributive property
           $= -3x$ Addition

5. $7x + 3x + 2x = (7 + 3 + 2)x$ Distributive property
                 $= 12x$

9. $4x - 3 + 2x = 4x + 2x - 3$ Commutative property
              $= (4x + 2x) - 3$ Associative property
              $= (4 + 2)x - 3$ Distributive property
              $= 6x - 3$

13. $2x - 3 + 3x - 2 = 2x + 3x - 3 - 2$ Commutative property
              $= (2x + 3x) + (-3 - 2)$ Associative property
              $= (2 + 3)x + (-3 - 2)$ Distributive property
              $= 5x - 5$ Addition

17. $-4x + 8 - 5x - 10 = -4x - 5x + 8 - 10$ Commutative property
              $= (-4x - 5x) + (8 - 10)$ Associative property
              $= (-4 - 5)x + 8 - 10)$ Distributive property
              $= -9x - 2$ Addition

21. $5(2x - 1) + 4 = 5(2x) - 5(1) + 4$ Distributive property
              $= 10x - 5 + 4$ Multiplication
              $= 10x - 1$ Addition

25. $-3(2x - 1) + 5 = 3(2x) - (-3)(1) + 5$ Distributive property
              $= -6x + 3 + 5$ Multiplication
              $= -6x + 8$ Addition

29. $6 - 4(x - 5) = 6 - 4(x) - (-4)(5)$ Distributive property
              $= 6 - 4x + 20$ Multiplication
              $= -4x + 6 + 20$ Commutative property
              $= -4x + 26$ Addition

33. $-6 + 2(2 - 3x) + 1 = -6 + 4 - 6x + 1$ Distributive property
              $= -6x - 6 + 4 + 1$ Commutative property
              $= -6x - 1$ Addition

37. $4(2y - 8) - (y + 7)$

Remember: $-(y + 7) = -1(y + 7) = -1(y) + (-1)(7) = -y - 7$

   $4(2y - 8) - (y + 7) = 8y - 32 - y - 7$ Distributive property
                       $= 8y - y - 32 - 7$ Commutative property
                       $= 7y - 39$ Add similar terms

Section 2.1 continued

41. When $x = 2$, $3x - 1 = 3(2) - 1$
$$= 6 - 1$$
$$= 5$$

45. When $x = 2$, $-2x - 5 = -2(2) - 5$
$$= -4 - 5$$
$$= -9$$

49. When $x = 2$, $(x - 4)^2 = (2 - 4)^2$
$$= (-2)^2$$
$$= 4$$

53. When $x = 2$, $(x + 3)(x - 3) = (2 + 3)(2 - 3)$
$$= (5)(-1)$$
$$= -5$$

57. When $x = -5$, $5x + 6 + 2x + 1 = 5(-5) + 6 + 2(-5) + 1$
$$= -25 + 6 + (-10) + 1$$
$$= -28$$

61. When $x = -5$, $5(2x + 1) + 4 = 5[2(-5) + 1] + 4$
$$= 5(-10 + 1) + 4$$
$$= 5(-9) + 4$$
$$= -45 + 4$$
$$= -41$$

65. When $x = -3$ and $y = 5$, $-2x - y - 9 = -2(-3) - (5) - 9$
$$= 6 - 5 - 9$$
$$= -8$$

69. When $x = -3$ and $y = 5$, $(x - y)^2 = (-3 - 5)^2$
$$= (-8)^2$$
$$= 64$$

73. When $x = -3$ and $y = 5$, $(x + 3y)^2 = [-3 + 3(5)]^2$
$$= (-3 + 15)^2$$
$$= (12)^2$$
$$= 144$$

77. When $x = \frac{1}{4}$, $12x - 3 = 12(\frac{1}{4}) - 3$
$$= 3 - 3 \qquad 12(\frac{1}{4}) = \frac{12}{4} = 3$$
$$= 0$$

Section 2.1 continued

81. When $x = \frac{3}{4}$, $12(x) - 3 = 12(\frac{3}{4}) - 3$

$$= 9 - 3 \qquad\qquad 12(\frac{3}{4}) = \frac{36}{4} = 9$$
$$= 6$$

85. When $x = -2$, $x - 5 = -2 - 5$
$$= -7$$

89. When $x = -2$, $\frac{10}{x} = \frac{10}{-2}$
$$= -5$$

93. $\frac{6}{8} - \frac{2}{8} = \frac{6 - 2}{8} = \frac{4}{8} = \frac{1}{2}$

97. $\frac{3}{x} - \frac{5}{6} = \frac{3 \cdot 6}{x \cdot 6} - \frac{5 \cdot x}{6 \cdot x}$ Change to equivalent fractions, LCD = 6x

$$= \frac{18}{6x} - \frac{5x}{6x}$$

$$= \frac{18 - 5x}{6x} \qquad\qquad \text{Add the numerator}$$

Section 2.2

1. $x - 3 = 8$
$x - 3 + \mathbf{3} = 8 + \mathbf{3}$ Add **3** to both sides
$x = 11$ Simplify both sides

5. $a + \frac{1}{2} = -\frac{1}{4}$

$a + \frac{1}{2} + (-\frac{1}{2}) = -\frac{1}{4} + (-\frac{1}{2})$ Add $-\frac{1}{2}$ to both sides

$a = -\frac{1}{4} + (-\frac{2}{4}) - \frac{1}{2} = -\frac{2}{4}$

$a = -\frac{3}{4}$ Simplify both sides

9. $y + 11 = -6$
$y + 11 + (\mathbf{-11}) = -6 + (\mathbf{-11})$ Add **-11** to both sides
$y = -17$ Simplify both sides

13.
$$m - 6 = -10$$
$$m - 6 + \mathbf{6} = -10 + \mathbf{6} \qquad \text{Add } \mathbf{6} \text{ to both sides}$$
$$m = -4$$

17.
$$5 = a + 4$$
$$5 + (\mathbf{-4}) = a + 4 + (\mathbf{-4}) \qquad \text{Add } \mathbf{-4} \text{ to both sides}$$
$$1 = a$$

21.
$$4x + 2 - 3x = 4 + 1$$
$$x + 2 = 5 \qquad \text{Simplify both sides first}$$
$$x + 2 + (\mathbf{-2}) = 5 + (\mathbf{-2}) \qquad \text{Add } \mathbf{-2} \text{ to both sides}$$
$$x = 3 \qquad \text{Simplify both sides}$$

25.
$$-3 - 4x + 5x = 18$$
$$-3 + x = 18 \qquad \text{Simplify the left side}$$
$$-3 + \mathbf{3} + x = 18 + \mathbf{3} \qquad \text{Add } \mathbf{3} \text{ to both sides}$$
$$x = 21 \qquad \text{Simplify both sides}$$

29.
$$-2.5 + 4.8 = 8x - 1.2 - 7x$$
$$2.3 = x - 1.2 \qquad \text{Simplify both sides first}$$
$$2.3 + \mathbf{1.2} = x - 1.2 + \mathbf{1.2} \qquad \text{Addition property of equality}$$
$$3.5 = x$$

33.
$$15 - 21 = 8x + 3x - 10x$$
$$-6 = x \qquad \text{Simplify both sides}$$

37.
$$2(x + 3) - x = 4$$
$$2x + 6 - x = 4 \qquad \text{Distributive property}$$
$$x + 6 = 4 \qquad \text{Simplify the left side}$$
$$x + 6 + (\mathbf{-6}) = 4 + (\mathbf{-6}) \qquad \text{Add } \mathbf{-6} \text{ to both sides}$$
$$x = -2 \qquad \text{Simplify both sides}$$

41.
$$5(2a + 1) - 9a = 8 - 6$$
$$10a + 5 - 9a = 8 - 6 \qquad \text{Distributive property}$$
$$a + 5 = 2 \qquad \text{Simplify both sides}$$
$$a + 5 + (\mathbf{-5}) = 2 + (\mathbf{-5}) \qquad \text{Add } \mathbf{-5} \text{ to both sides}$$
$$a = -3 \qquad \text{Simplify both sides}$$

45.
$$4y - 3(y - 6) + 2 = 8$$
$$4y - 3y + 18 + 2 = 8 \qquad \text{Distributive property}$$
$$y + 20 = 8 \qquad \text{Simplify the left side}$$
$$y + 20 + (\mathbf{-20}) = 8 + (\mathbf{-20}) \qquad \text{Add } \mathbf{-20} \text{ to both sides}$$
$$y = -12 \qquad \text{Simplify both sides}$$

49.
$$-3(2m - 9) + 7(m - 4) = 12 - 9$$
$$-6m + 27 + 7m - 28 = 12 - 9 \qquad \text{Distributive property}$$
$$m - 1 = 3 \qquad \text{Simplify both sides}$$
$$m - 1 + \mathbf{1} = 3 + \mathbf{1} \qquad \text{Add } \mathbf{1} \text{ to both sides}$$
$$m = 4 \qquad \text{Simplify both sides}$$

Section 2.2 continued

53.
$$8a = 7a - 5$$
$$8a + (-7a) = 7a + (-7a) - 5 \qquad \text{Add } -7a \text{ to both sides}$$
$$a = -5 \qquad\qquad\qquad \text{Simplify both sides}$$

57.
$$3y + 4 = 2y + 1$$
$$3y + (-2y) + 4 = 2y + (-2y) + 1 \qquad \text{Add } -2y \text{ to both sides}$$
$$y + 4 = 1 \qquad\qquad\qquad \text{Simplify both sides}$$
$$y + 4 + (-4) = 1 + (-4) \qquad \text{Add } -4 \text{ to both sides}$$
$$y = -3 \qquad\qquad\qquad \text{Simplify both sides}$$

61.
$$4x - 7 = 5x + 1$$
$$4x + (-4x) - 7 = 5x + (-4x) + 1 \qquad \text{Add } -4x \text{ to both sides}$$
$$-7 = x + 1 \qquad\qquad\qquad \text{Simplify both sides}$$
$$-7 + (-1) = x + 1 + (-1) \qquad \text{Add } -1 \text{ to both sides}$$
$$-8 = x \qquad\qquad\qquad \text{Simplify both sides}$$

65.
$$8a - 7.1 = 7a + 3.9$$
$$8a + (-7a) - 7.1 = 7a + (-7a) + 3.9 \qquad \text{Add } (-7a) \text{ to both sides}$$
$$a - 7.1 = 3.9$$
$$a - 7.1 + 7.1 = 3.9 + 7.1 \qquad \text{Add } 7.1 \text{ to both sides}$$
$$a = 11.0$$
$$a = 11$$

71.
$$8\left(\tfrac{1}{8}y\right) = \left(\tfrac{8}{1} \cdot \tfrac{1}{8}\right)y \qquad \text{Associative property}$$
$$= 1y \qquad\qquad \text{Reciprocals}$$
$$= y$$

75.
$$-\tfrac{4}{3}\left(-\tfrac{3}{4}a\right) = \left[-\tfrac{4}{3}\left(-\tfrac{3}{4}\right)\right]a \qquad \text{Associative property}$$
$$= 1a \qquad\qquad \text{Reciprocals}$$
$$= a$$

Section 2.3

1.
$$5x = 10$$
$$\tfrac{1}{5}(5x) = \tfrac{1}{5}(10) \qquad \text{Multiply both sides by } \tfrac{1}{5}$$
$$\left[\tfrac{1}{5}(5)\right]x = \tfrac{1}{5}(10) \qquad \text{Associative property}$$
$$x = 2 \qquad\qquad \text{Simplify: } \tfrac{1}{5}(10) = \tfrac{1}{5}\left(\tfrac{10}{1}\right) = \tfrac{10}{5} = 2$$

5. $\qquad -8x = 4$

$\qquad -\dfrac{1}{8}(-8x) = -\dfrac{1}{8}(4) \qquad$ Multiply both sides by $-\dfrac{1}{8}$

$\qquad \left[-\dfrac{1}{8}(-8)\right]x = -\dfrac{1}{8}(4) \qquad$ Associative property

$\qquad x = -\dfrac{1}{2} \qquad$ Simplify: $\quad -\dfrac{1}{8}(4) = -\dfrac{1}{8}\left(\dfrac{4}{1}\right) = -\dfrac{4}{8} = -\dfrac{1}{2}$

9. $\qquad -3x = -9$

$\qquad -\dfrac{1}{3}(-3x) = -\dfrac{1}{3}(-9) \qquad$ Multiply both sides by $-\dfrac{1}{3}$

$\qquad x = 3$

13. $\qquad 2x = 0$

$\qquad \dfrac{1}{2}(2x) = \dfrac{1}{2}(0) \qquad$ Multiply both sides by $\dfrac{1}{2}$

$\qquad x = 0$

17. $\qquad \dfrac{x}{3} = 2 \qquad$ Dividing by 3 is equivalent to multiplying by $\dfrac{1}{3}$

$\qquad \dfrac{1}{3}x = 2$

$\qquad 3\left(\dfrac{1}{3}x\right) = 3(2) \qquad$ Multiply both sides by **3**

$\qquad x = 6$

21. $\qquad -\dfrac{x}{2} = -\dfrac{3}{4}$

$\qquad -\dfrac{1}{2}x = -\dfrac{3}{4} \qquad$ Dividing by 2 is equivalent to multiplying by $\dfrac{1}{2}$

$\qquad -2\left(-\dfrac{1}{2}x\right) = -2\left(-\dfrac{3}{4}\right) \qquad$ Multiply both sides by **-2**

$\qquad x = \dfrac{6}{4}$

$\qquad x = \dfrac{3}{2} \qquad$ Reduce to lowest terms

25.
$$-\frac{3}{5}x = \frac{9}{5}$$

$$-\frac{5}{3}\left(-\frac{3}{5}x\right) = -\frac{5}{3}\left(\frac{9}{5}\right) \qquad \text{Multiply both sides by } -\frac{5}{3}$$

$$x = \frac{-5}{3} \cdot \frac{9}{5}$$

$$x = -\frac{45}{15}$$

$$x = -3$$

29.
$$-4x - 2x + 3x = 24$$
$$-3x = 24 \qquad \text{Simplify the left side}$$

$$-\frac{1}{3}(3x) = -\frac{1}{3}(24) \qquad \text{Multiply both sides by } -\frac{1}{3}$$

$$x = -\frac{1}{3}\left(\frac{24}{1}\right)$$

$$x = \frac{-24}{3}$$

$$x = -8$$

33.
$$-3 - 5 = 3x + 5x - 10x$$
$$-8 = 2x \qquad \text{Simplify both sides}$$

$$-\frac{1}{2}(-8) = -\frac{1}{2}(-2x) \qquad \text{Multiply both sides by } -\frac{1}{2}$$

$$4 = x$$

37.
$$-x = 4$$
$$-1(-x) = -1(4) \qquad \text{Multiply both sides by } -1$$
$$x = -4$$

41.
$$15 = -a$$
$$-1(15) = -1(-a) \qquad \text{Multiply both sides by } -1$$
$$-15 = a$$

45.
$$3x - 2 = 7$$
$$3x - 2 + 2 = 7 + 2 \qquad \text{Add 2 to both sides}$$
$$3x = 9 \qquad \text{Simplify}$$

$$\frac{1}{3}(3x) = \frac{1}{3}(9) \qquad \text{Multiply both sides by } \frac{1}{3}$$

$$x = 3$$

49. Method 1

$$\frac{1}{8} + \frac{1}{2}x = \frac{1}{4}$$

$$\frac{1}{8} + \left(-\frac{1}{8}\right) + \frac{1}{2}x = \frac{1}{4} + \left(-\frac{1}{8}\right) \qquad \text{Add } -\frac{1}{8} \text{ to each side}$$

$$\frac{1}{2}x = \frac{1}{8} \qquad\qquad \frac{1}{4} + \left(-\frac{1}{8}\right) = \frac{2}{8} + \left(-\frac{1}{8}\right) = \frac{1}{8}$$

$$\frac{2}{1}\left(\frac{1}{2}x\right) = \frac{2}{1}\left(\frac{1}{8}\right) \qquad \text{Multiply each side by } \frac{2}{1}$$

$$x = \frac{2}{8}$$

$$x = \frac{1}{4}$$

Method 2

$$8\left(\frac{1}{8} + \frac{1}{2}x\right) = 8\left(\frac{1}{4}\right) \qquad \text{Multiply each side by the LCD } 8$$

$$8\left(\frac{1}{8}\right) + 8\left(\frac{1}{2}x\right) = 8\left(\frac{1}{4}\right) \qquad \text{Distribute property on the left side}$$

$$1 + 4x = 2 \qquad\qquad \text{Multiply}$$
$$1 + (-1) + 4x = 2 + (-1) \qquad \text{Add } -1 \text{ to each side}$$
$$4x = 1$$

$$\frac{1}{4}(4x) = \frac{1}{4}(1) \qquad \text{Multiply each side by } \frac{1}{4}$$

$$x = \frac{1}{4}$$

53.
$$2y = -4y + 18$$
$$2y + 4y = -4y + 4y + 18 \qquad \text{Add } 4y \text{ to both sides}$$
$$6y = 18 \qquad\qquad \text{Simplify}$$

$$\frac{1}{6}(6y) = \frac{1}{6}(18) \qquad \text{Multiply both sides by } \frac{1}{6}$$

$$y = 3$$

57.
$$8x + 4 = 2x - 5$$
$$8x + (-2x) + 4 = 2x + (-2x) - 5 \qquad \text{Add } -2x \text{ to both sides}$$
$$6x + 4 = -5$$
$$6x + 4 + (-4) = -5 + (-4) \qquad \text{Add } -4 \text{ to both sides}$$
$$6x = -9$$

$$\frac{1}{6}(6x) = \frac{1}{6}(-9) \qquad \text{Multiply both sides by } \frac{1}{6}$$

$$x = -\frac{9}{6}$$

$$x = -\frac{3}{2} \qquad \text{Reduce to lowest terms}$$

Section 2.3 continued

61.
$$6m - 3 = m + 2$$
$$6m + (-m) - 3 = m + (-m) + 2 \qquad \text{Add } -m \text{ to each side}$$
$$5m - 3 = 2$$
$$5m - 3 + 3 = 2 + 3 \qquad \text{Add } 3 \text{ to each side}$$
$$5m = 5 \qquad \text{Add}$$
$$\frac{1}{5}(5m) = \frac{1}{5}(5) \qquad \text{Multiply both sides by } \frac{1}{5}$$
$$m = 1$$

65.
$$9y + 2 = 6y - 4$$
$$9y + (-6y) + 2 = 6y + (-6y) - 4 \qquad \text{Add } -6y \text{ to each side}$$
$$3y + 2 = -4$$
$$3y + 2 + (-2) = -4 + (-2) \qquad \text{Add } -2 \text{ to each side}$$
$$3y = -6$$
$$\frac{1}{3}(3y) = \frac{1}{3}(-6) \qquad \text{Multiply both sides by } \frac{1}{3}$$
$$y = -2$$

69.
$$2x - 5 = 2x + 3$$
$$2x + (-2x) - 5 = 2x + (-2x) + 3 \qquad \text{Add } -2x \text{ to each side}$$
$$-5 = 3 \qquad \text{False statement}$$

73.
$$5(2x - 8) - 3 = 5(2x) + 5(-8) = -3 \qquad \text{Distributive property}$$
$$= 10x - 40 - 3 \qquad \text{Multiply}$$
$$= 10x - 43$$

77.
$$7 - 3(2y + 1) = 7 + (-3)(2y) + (-3)(1) \qquad \text{Distributive property}$$
$$= 7 - 6y - 3 \qquad \text{Multiply}$$
$$= -6y + 4 \qquad \text{Simplify}$$

Section 2.4

1. Solve: $2(x + 3) = 12$

Solution: Our first step is to apply the distributive property to the left side of the equation:

$$2x + 6 = 12 \qquad \text{Distributive property}$$
$$2x + 6 + (-6) = 12 + (-6) \qquad \text{Add } -6 \text{ to both sides}$$
$$2x = 6 \qquad \text{Simplify both sides}$$
$$\frac{1}{2}(2x) = \frac{1}{2}(6) \qquad \text{Multiply both sides by } \frac{1}{2}$$
$$x = 3$$

5. Solve: $2(4a + 1) = -6$

Solution: Distribute the 2 across the sum of $4a + 1$:

$$8a + 2 = -6 \qquad \text{Distributive property}$$
$$8a + 2 + (-2) = -6 + (-2) \qquad \text{Add } -2 \text{ to both sides}$$
$$8a = -8 \qquad \text{Simplify both sides}$$

$$\frac{1}{8}(8a) = \frac{1}{8}(-8) \qquad \text{Multiply both sides by } \frac{1}{8}$$

$$a = -1$$

9. Solve: $-2(3y + 5) = 14$

Solution: We begin by multiplying -2 times the sum of $3y + 5$:

$$-6y - 10 = 14 \qquad \text{Distributive property}$$
$$-6y - 10 + 10 = 14 + 10 \qquad \text{Add } 10 \text{ to both sides}$$
$$-6y = 24 \qquad \text{Simplify both sides}$$

$$* \qquad -\frac{1}{6}(-6y) = -\frac{1}{6}(24) \qquad \text{Multiply both sides by } -\frac{1}{6}$$

$$y = -4$$

$*$ Remember to multiply by the same sign as the coefficient so the variable will have a coefficient of positive one when you have completed the problem.

13. $1 = \frac{1}{2}(4x + 2)$

$$1 = \frac{1}{2}(4x) + \frac{1}{2}(2) \qquad \text{Distributive property}$$

$$1 = 2x + 1 \qquad \text{Multiply}$$
$$0 = 2x \qquad \text{Add } -1 \text{ to each side}$$

$$0 = x \qquad \text{Multiply both sides by } \frac{1}{2}$$

17. Solve: $4(2y + 1) - 7 = 1$

Solution: Distribute the 4 across the sum $2y + 1$

$$8y + 4 - 7 = 1 \qquad \text{Distributive property}$$
$$8y - 3 = 1 \qquad \text{Simplify the left side}$$
$$8y - 3 + 3 = 1 + 3 \qquad \text{Add } + 3 \text{ to both sides}$$
$$8y = 4 \qquad \text{Simplify both sides}$$

$$\frac{1}{8}(8y) = \frac{1}{8}(4) \qquad \text{Multiply both sides by } \frac{1}{8}$$

$$y = 2$$

Remember to check the problems if you think it is necessary.

21. Solve: $-7(2x - 7) = 3(11 - 4x)$

Solution: We begin by removing the parentheses by multiplying:

$-7(2x - 7) = 3(11 - 4x)$	Original equation
$-14x + 49 = 33 - 12x$	Distributive property
$-14x + \mathbf{14x} + 49 = 33 - 12x + \mathbf{14x}$	Add **14x** to both sides
$49 = 33 + 2x$	Simplify both sides
$49 + (\mathbf{-33}) = 33 + (\mathbf{-33}) + 2x$	Add **-33** to both sides
$16 = 2x$	
$\frac{1}{2}(16) = \frac{1}{2}(2x)$	Multiply both sides by $\frac{1}{2}$
$8 = x$	

25.
$$\frac{3}{4}(8x - 4) + 3 = \frac{2}{5}(5x + 10) - 1$$

$\frac{3}{4}(8x) + \frac{3}{4}(-4) + 3 = \frac{2}{5}(5x) + \frac{2}{5}(10) - 1$	Distributive property
$6x - 3 + 3 = 2x + 4 - 1$	Multiply
$6x = 2x + 3$	Simplify
$4x = 3$	Add -2x to each side
$x = \frac{3}{4}$	Multiply both sides by $\frac{1}{4}$

29. Solve $6 - 5(2a - 3) = 1$

Solution: Begin by multiplying -5 times the difference of 2a - 3:

$6 - 10a + 15 = 1$	Distributive property
$10a + 21 = 1$	Simplify the left side
$-10a + 21 + (\mathbf{-21}) = 1 + (\mathbf{-21})$	Add **-21** to both sides
$-10a = -20$	Simplify both sides
$-\frac{1}{10}(-10a) = -\frac{1}{10}(-20)$	Multiply both sides by $-\frac{1}{10}$
$a = 2$	

33. Solve: $2(t - 3) + 3(t - 2) = 28$

Solution: Begin by applying the distributive property to each parentheses:

$2(t - 3) + 3(t - 2) = 28$	Original equation
$2t - 6 + 3t - 6 = 28$	Distributive property
$5t - 12 = 28$	Simplify the left side
$5t - 12 + \mathbf{12} = 28 + \mathbf{12}$	Add **12** to both sides
$5t = 40$	Simplify each side
$\frac{1}{5}(5t) = \frac{1}{5}(40)$	Multiply both sides by $\frac{1}{5}$
$t = 8$	

37. Solve: $2(5x - 3) - (2x - 4) = 5 - (6x + 1)$

Solution: When we apply the distributive property, we have to be careful with signs. Remember, we can think of $-(6x + 1)$ as $-1(6x + 1)$ so that $-(6x + 1) = -1(6x + 1) = -6x - 1$.

$2(5x - 3) - (2x - 4) = 5 - (6x + 1)$	Original equation
$10x - 6 - 2x + 8 = 5 - 6x - 1$	Distributive property
$8x + 2 = -6x + 4$	Simplify both sides
$8x + \mathbf{6x} + 2 = -6x + \mathbf{6x} + 4$	Add **6x** to both sides
$14x + 2 = 4$	Simplify both sides
$14x + 2 + (\mathbf{-2}) = 4 + (\mathbf{-2})$	Add **-2** to both sides
$14x = 2$	Simplify both sides
$\frac{1}{14}(14x) = \frac{1}{14}(2)$	Multiply both sides by $\frac{1}{14}$
$x = \frac{2}{14}$	Simplify the right side
$x = \frac{1}{7}$	Reduce to lowest terms

41. Solve: $2(x + 3) = 2x + 6$

Solution: Begin by applying the distributive property to the parentheses:

$2x + 6 = 2x + 6$	Distributive property
$2x + (\mathbf{-2x}) + 6 = 2x + (\mathbf{-2x}) + 6$	Add **-2x** to both sides
$6 = 6$	True statement

Solution: All real numbers.

45. No. 47 was done instead of No. 45 to show a different solution.

47. Solve: $4x + 6 = 2(2x - 6)$

Solution: We begin by multiplying 2 times the difference of $2x - 6$:

$4x + 6 = 4x - 6$	Distributive property
$4x + (\mathbf{-4x}) + 6 = 4x + (\mathbf{-4x}) - 6$	Add **-4x** to both sides
$6 = -6$	False statement

There is no solution to the equation.

51. $\frac{2}{3}(6) = \frac{2}{3}\left(\frac{6}{1}\right)$

$= \frac{12}{3}$

$= 4$

57. $\frac{1}{2}(3x - 6) = \frac{1}{2}(3x) - \frac{1}{2}(6)$

$= \frac{3}{2}x - 3$

Section 2.5

1. Substituting P = 300 and W = 50 into P = 2L + 2W, we have

$$300 = 2L + 2(50)$$
$$300 = 2L + 100$$

Now we solve for L

$200 = 2L$	Add -100 to both sides
$\frac{1}{2}(200) = \frac{1}{2}(2L)$	Multiply by $\frac{1}{2}$
$100 = L$	

The length is 100 feet.

5. Substituting x = 0 into 2x + 3y = 6, we have

$$2(0) + 3y = 6$$
$$3y = 6$$

$\frac{1}{3}(3y) = \frac{1}{3}(6)$	Multiply by $\frac{1}{3}$
$y = 2$	

9. Substituting y = 0 into 2x - 5y = 20, we have

$$2x - 5(0) = 20$$
$$2x = 20$$
$$\frac{1}{2}(2x) = \frac{1}{2}(20)$$
$$x = 10$$

13. Substituting y = 3 into y = 2x - 1, gives us

$$3 = 2x - 1$$
$$3 + 1 = 2x - 1 + 1$$
$$4 = 2x$$
$$\frac{1}{2}(4) = \frac{1}{2}(2x)$$
$$2 = x$$

17. Solve: d = rt for r

Solution: $\frac{1}{t}(d) = r(\frac{1}{t})t$	Multiply both sides by $\frac{1}{t}$
$\frac{d}{t} = r$	

Section 2.5 continued

21. $PV = nRT$

$$\frac{PV}{V} = \frac{nRT}{V} \qquad \text{Divide both sides by V}$$

$$P = \frac{nRT}{V}$$

25. Solve: $P = a + b + c$ for a

Solution:
$$P + (-b) = a + b + (-b) + c$$
$$P - b = a + c$$
$$P - b + (-c) = a + c + (-c)$$
$$P - b - c = a$$

This may be done in one step by adding $-b + (-c)$ to both sides

29. Solve: $-3x + y = 6$ for y

Solution: $\quad -3x + 3x + y = 6 + 3x \qquad$ Add **3x** to both sides
$$y = 3x + 6$$

33. Solve: $6x + 3y = 12$ for y

Solution: $\quad 6x + (-6x) + 3y = 12 + (-6x) \qquad$ Add **-6x** to both sides
$$3y = -6x + 12$$

$$\frac{1}{3}(3y) = \frac{1}{3}(-6x + 12) \qquad \text{Multiply both sides by } \frac{1}{3}$$

$$y = \frac{1}{3}(-6x) + \frac{1}{3}(12) \quad \text{Distributive property}$$

$$y = -2x + 4 \qquad \text{Multiplication}$$

37. Solve: $P = 2L + 2W$ for W

Solution: $\quad P + (-2L) = 2L + (-2L) + 2W \qquad$ Add **-2L** to both sides
$$P - 2L = 2W$$

$$\frac{1}{2}(P - 2L) = \frac{1}{2}(2W) \qquad \text{Multiply both sides by } \frac{1}{2}$$

$$\frac{P - 2L}{2} = W \quad \text{or} \quad W = \frac{P}{2} - L$$

41. Solve: $h = vt + 16t^2$ for v

Solution: $\quad h + (-16t^2) = vt + 16t^2 + (-16t^2) \quad$ Add **-16t²** to both sides
$$y - 16t^2 = vt$$

$$\frac{1}{t}(h - 16t^2) = \frac{1}{t}(vt) \qquad \text{Multiply both sides by } \frac{1}{t}$$

$$\frac{h - 16t^2}{t} = v$$

45. Method 1

$$\frac{x}{2} + \frac{y}{3} = 1$$

$$\frac{y}{3} = -\frac{x}{2} + 1 \qquad \text{Add } -\frac{x}{2} \text{ to each side}$$

$$3\left(\frac{y}{3}\right) = 3\left(-\frac{x}{2} + 1\right) \qquad \text{Multiply each side by 3}$$

$$y = -\frac{3}{2}x + 3 \qquad \text{Distributive property}$$

Method 2

$$6\left(\frac{x}{2}\right) + 6\left(\frac{y}{3}\right) = 6(1) \qquad \text{Multiply both sides by LCD} = 6$$

$$3x + 2y = 6$$
$$2y = -3x + 6 \qquad \text{Add } -3x \text{ to both sides}$$

$$y = \frac{-3x}{2} + \frac{6}{2} \qquad \text{Multiply both sides by } \frac{1}{2}$$

$$y = -\frac{3}{2}x + 3 \qquad \text{Simplify}$$

49. Method 1

$$-\frac{1}{4}y + \frac{1}{8}y = 1$$

$$\frac{1}{8}y = \frac{1}{4}x + 1 \qquad \text{Add } \frac{1}{4}x \text{ to both sides}$$

$$8\left(\frac{1}{8}y\right) = 8\left(\frac{1}{4}x\right) + 8(1) \qquad \text{Multiply both sides by 8}$$

$$y = 2x + 8$$

Method 2

$$8\left(-\frac{1}{4}x\right) + 8\left(\frac{1}{8}y\right) = 8(1) \qquad \text{Multiply both sides by LCD} = 8$$

$$-2x + y = 8$$
$$y = 2x + 8 \qquad \text{Add } 2x \text{ to each side}$$

Section 2.5 continued

53. If $C = 9.42$, $\pi = 3.14$ and $ = 2\pi r$; then

$$C = 2\pi r$$
$$9.42 = 2(3.14)r \qquad \text{Substitution}$$
$$9.42 = 6.28r \qquad \text{Multiply}$$

$$\frac{9.42}{6.28} = \frac{6.28r}{6.28} \qquad \text{Divide each side by 6.28}$$

$$\frac{3}{2} = r$$

The r (radius) equals $\frac{3}{2}$ inches or 1.5 inches.

57. If $V = 6.28$, $r = 3$, and $\pi = 3.14$, then

$$V = \pi r^2 h$$
$$6.28 = (3.14)(3^2)h \qquad \text{Substitution}$$
$$6.28 = (3.14)(9)h$$
$$6.28 = 28.26h$$

$$\frac{6.28}{28.26} = \frac{28.26h}{28.26h} \qquad \text{Divide by 28.26 to each side}$$
$$\frac{2}{9} = h$$

The h (height) is $\frac{2}{9}$ centimeters.

61. Substituting $r = 4$, $h = 11$ and $\pi = 3.14$ into $A = 2\pi r^2 + 2\pi rh$, we have

$$A = 2(3.14)(4^2) + 2(3.14)(4)11$$
$$A = 2(3.14)(16) + 2(3.14)(4)11 \qquad \text{Simplify exponents}$$
$$A = 6.28(16) + 6.28(4)11$$
$$A = 100.48 + 25.12(11)$$
$$A = 100.48 + 276.32$$
$$A = 376.80 \text{ square centimeters}$$

65. What number is 12% of 2000?

$$N = .12 \cdot 2000$$
$$N = 240.00$$
$$N = 240$$

69. What percent of 40 is 14?

$$N \cdot 40 = 14$$
$$40N = 14$$

$$\frac{40N}{40} = \frac{14}{40}$$

$$N = \frac{7}{20} = .35 = 35\%$$

Section 2.5 continued

73. 240 is 12% of what number?

$$240 = .12 \cdot x$$
$$240 = .12x$$

$$\frac{240}{.12} = \frac{.12x}{.12}$$

$$2000 = x$$

77. The difference of 6 and 2 is 4.

81. $2 \cdot 5 + 3 = 13$ Twice means two times.

Section 2.6

1. Step 1: Let x = the number asked for

Step 2: The sum of a number and five is thirteen

$$x + 5 = 13$$

Step 3:
$$x + 5 = 13$$
$$x + 5 + (-5) = 13 + (-5)$$
$$x = 8$$

Step 4:
$$x + 5 = 13$$
$$8 + 5 = 13 \quad \text{True}$$

5. Step 1: Let x = the number asked for

Step 2: Five times the sum of a number and seven is thirty

$$5(x + 7) = 30$$

Step 3:
$$5(x + 7) = 30$$
$$5x + 35 = 30$$
$$5x + 35 + (-35) = 30 \, (-35)$$
$$5x = -5$$
$$x = -1$$

Step 4:
$$5(x + 7) = 30$$
$$5(-1 + 7) = 30$$
$$5(6) = 30$$
$$30 = 30$$

9. 1st number = x
2nd number = 3x - 4

$$(x + 3x - 4) + 5 = 25$$
$$4x - 4 + 5 = 25$$
$$4x + 1 = 25$$
$$4x = 24$$
$$x = 6$$

The first number is x = 6 and the second number is
$3x - 4 = 3(6) - 4 = 14$.

13.

	Now	In three years
Jack	$2x$	$2x + 3$
Lacy	x	$x + 3$

$$(2x + 3) + (x + 3) = 54$$
$$3x + 6 = 54$$
$$3x = 48$$
$$x = 16$$

Lacy is $x = 16$ years old. Jack is $2x = 2(16) = 32$ years old.

17. Length $= x + 5$
Width $= x$
Perimeter $= 34$ inches

$$2 \text{ lengths} + 2 \text{ widths} = \text{perimeter}$$
$$2(x + 5) + 2x = 34 \text{ inches}$$
$$2x + 10 + 2x = 34$$
$$4x + 10 = 34$$
$$4x = 24$$
$$x = 6$$

The width is $x = 6$ inches. The length is $x + 5 = 6 + 5 = 11$ inches.

21. Width $= x$
Length $= 2x - 3$
Perimeter $= 54$ inches

$$2 \text{ lengths} + 2 \text{ widths} = 54 \text{ inches}$$
$$2(2x - 3) + 2x = 54$$
$$4x - 6 + 2x = 54$$
$$6x - 6 = 54$$
$$6x - 6 + 6 = 54 + 6$$
$$6x = 60$$
$$x = 10$$

The width is $x = 10$ inches. The length is $2x - 3 = 2(10) - 3 = 17$ inches.

25.

	Dimes	Quarter
Number	x	$2x$
Value (in cents)	$10x$	$25(2x)$

$$\text{Amount of money in dimes} + \text{Amount of money in quarters} = \text{Total amount of money}$$
$$10x + 25(2x) = 900*$$

$$10x + 50x = 900 \quad \text{Multiply}$$
$$60x = 900 \quad \text{Simplify the left side}$$
$$x = 15 \quad \text{Divide each side by 60}$$

You have $x = 15$ dimes and $2x = 2(15) = 30$ quarters.

*Remember $9.00 equals 900 cents.

29.

	Dollars invested at 8%	Dollars invested at 9%
Number of	x	x + 2000
Interest on	.08x	.09(x + 2000)

Interest earned at 8%	+	Interest earned at 9%	=	Total interest earned
.08x	+	.09(x + 2000)	=	860

.08x + .09x + 180 = 860	Distributive property
.17x + 180 = 860	Add similar terms
.17x = 680	Add -180 to each side
x = 4000	Divide each side by 17

The amount of money invested at 8% is x = $4000 and at 9% is x + $2000 = $6000.

33.

	Dollars invested at 8%	Dollars invested at 9%	Dollars invested at 10%
Number of	x	2x	3x
Interest on	.08x	.09(2x)	.10(3x)

Interest earned at 8%	+	Interested earned at 9%	+	Interest earned at 10%	=	Total interest earned
.08x	+	.09(2x)	+	.10(3x)	=	280

.08x + .18x + .30x = 280	Multiply
.56x = 280	Add similar terms
x = 500	Divide each side by .5

The amount of money invested at 8% is x = $500, at 9% is 2x = $1000 and at 10% is 3x = $1500.

37. Let x = the number of hours past 35 hours.

35(12) + 18x = 492	
420 + 18x = 492	Multiply
18x = 72	
x = 4	Divide both sides by 18

She worked 35 hours plus 4 hours over-time so she worked 39 hours altogether.

41. 4 is less than 10

45. 12 < 20

Section 2.6 continued

49. $|8 - 3| - |5 - 2| = |5| - |3|$ Subtract

$$= 5 - 3 \qquad \text{Absolute value}$$
$$= 2 \qquad \text{Subtract}$$

Section 2.7   The graphs of the answers are shown in the textbook.

1.
$$x - 5 < 7$$
$$x - 5 + \mathbf{5} < 7 + \mathbf{5} \qquad \text{Add } \mathbf{5} \text{ to both sides}$$
$$x < 12$$

5.
$$x - 4.3 > 8.7$$
$$x - 4.3 + \mathbf{4.3} > 8.7 + \mathbf{4.3} \qquad \text{Add } \mathbf{4.3} \text{ to both sides}$$
$$x > 13.0$$

9.
$$2 < x - 7$$
$$2 + \mathbf{7} < x - 7 + \mathbf{7} \qquad \text{Add } \mathbf{7} \text{ to both sides}$$
$$9 < x \qquad \text{(Remember, this inequality may also be read}$$
$$x > 9)$$

13.
$$5a \le 25$$
$$\frac{1}{5}(5a) \le \frac{1}{5}(25) \qquad \text{Multiply each side by } \frac{1}{5}$$
$$a \le 5$$

17.
$$-2x > 6$$
$$-\frac{1}{2}(-2x) < -\frac{1}{2}(6) \qquad \text{Multiply each side by } -\frac{1}{2} \text{ and reverse the}$$
$$\qquad\qquad\qquad\qquad\qquad \text{direction of the inequality symbol}$$
$$x < -3$$

21.
$$-\frac{x}{5} \le 10$$
$$-5(-\frac{x}{5}) \ge -5(10) \qquad \text{Multiply each side by } \mathbf{-5} \text{ and reverse the direction}$$
$$\qquad\qquad\qquad\qquad\qquad \text{of the inequality symbol}$$
$$x \ge -50$$

25.
$$2x - 3 < 9$$
$$2x - 3 + \mathbf{3} < 9 + \mathbf{3} \qquad \text{Add } \mathbf{3} \text{ to both sides}$$
$$2x < 12$$
$$\frac{1}{2}(2x) < \frac{1}{2}(12) \qquad \text{Multiply each side by } \frac{1}{2}$$
$$x < 6$$

29.
$$-4x + 1 > -11$$
$$-4x + 1 - \mathbf{1} > -11 - \mathbf{1}$$ Add **-1** to both sides
$$-4x > -12$$

$$-\frac{1}{4}(-4x) < -\frac{1}{4}(-12)$$ Multiply each side by $-\frac{1}{4}$ and reverse the direction of the inequality symbol

$$x < 3$$

33.
$$-\frac{2}{5}a - 3 > 5$$

$$-\frac{2}{5}a - 3 + \mathbf{3} > 5 + \mathbf{3}$$ Add **3** to both sides

$$-\frac{2}{5}a > 8$$

$$-\frac{5}{2}(-\frac{2}{5}a) < -\frac{5}{2}(8)$$ Multiply each side by $-\frac{5}{2}$ and reverse the direction of the inequality symbol

$$a < -20$$

37.
$$.3(a + 1) \leq 1.2$$

$$.3a + .3 \leq 1.2$$ Distributive property

$$.3a + 3 - \mathbf{.3} \leq 1.2 - \mathbf{.3}$$ Add **-.3** to both sides

$$\frac{1}{.3}(.3a) \leq \frac{1}{.3}(.9)$$ Multiply each side by $\frac{1}{.3}$

$$a \leq 3$$

41.
$$3x - 5 > 8x$$
$$3x - \mathbf{3x} - 5 > 8x - \mathbf{3x}$$ Add **-3x** to both sides
$$-5 > 5x$$

$$\frac{1}{5}(-5) > \frac{1}{5}(5x)$$ Multiply each side by $\frac{1}{5}$

$$-1 > x$$ Remember this can also be read $x < -1$

Section 2.7 continued

45.   Method 1

$$-.4x + 1.2 < -2x - .4$$
$$-.4x + \textbf{2x} + 1.2 < -2x + \textbf{2x} - .4 \qquad \text{Add } \textbf{2x} \text{ to each side}$$
$$1.6x + 1.2 < .4$$
$$1.6x + 1.2 + (\textbf{-1.2}) < -.4 + (\textbf{-1.2}) \qquad \text{Add } \textbf{-1.2} \text{ to each side}$$
$$1.6x < -1.6$$
$$\frac{1.6x}{1.6} < \frac{-1.6}{1.6} \qquad \text{Divide by } 1.6$$
$$x < -1$$

Method 2

$$\textbf{10}(-.4x + 1.2) < \textbf{10}(-2x - .4) \qquad \text{Multiply each side by } \textbf{10}$$
$$\textbf{10}(-.4x) + \textbf{10}(1.2) < \textbf{10}(-2x) + \textbf{10}(-.4) \qquad \text{Distributive property}$$
$$-4x + 12 < -20x - 4$$
$$-4x + \textbf{20x} + 12 < -20x + \textbf{20x} - 4 \qquad \text{Add } \textbf{20x} \text{ to each side}$$
$$16x + 12 < -4$$
$$16x + 12 + (\textbf{-12}) < -4 + (\textbf{-12}) \qquad \text{Add } \textbf{-12} \text{ to each side}$$
$$16x < -16$$
$$\frac{16x}{\textbf{16}} < \frac{-16}{\textbf{16}} \qquad \text{Divide each side by } \textbf{16}$$
$$x < -1$$

See the graph on page A-6 in the textbook.

49. $3 - 4(x - 2) \leq -5x + 6$

$$3 - 4x + 8 \leq -5x + 6 \qquad \text{Distributive property}$$
$$-4x + 11 \leq -5x + 6 \qquad \text{Simplify the left side}$$
$$-4x + 11 - \textbf{11} \leq -5x + 6 - \textbf{11} \qquad \text{Add } \textbf{-11} \text{ to both sides}$$
$$-4x \leq -5x - 5$$
$$-4x + \textbf{5x} \leq -5x + \textbf{5x} - 5 \qquad \text{Add } \textbf{5x} \text{ to both sides}$$
$$x \leq -5$$

53. $2x - 5y > 10$

$$2x - \textbf{2x} - 5y > \textbf{-2x} + 10 \qquad \text{Add } \textbf{2x} \text{ to both sides}$$
$$-5y > -2x + 10$$
$$-\frac{1}{5}(-5y) < -\frac{1}{5}(-2x + 10) \qquad \text{Multiply each side by } -\frac{1}{5} \text{ and reverse}$$
$$\qquad\qquad\qquad\qquad\qquad\qquad \text{the direction of the inequality symbol}$$
$$y < \frac{2}{5}x - 2 \qquad \text{Distributive property}$$

Section 2.7 continued

57.
$$2x - 4y \geq -4$$

$$2x - 2x - 4y \geq -2x - 4 \qquad \text{Add } -2x \text{ to both sides}$$

$$-4y \geq -2x - 4$$

$$-\frac{1}{4}(4y) \leq -\frac{1}{4}(-2x - 4) \qquad \text{Multiply each side by } -\frac{1}{4} \text{ and reverse}$$
$$\text{the direction of the inequality symbol.}$$

$$y \leq \frac{1}{2}x + 1 \qquad \text{Distributive property}$$

61.
$$4x > x - 8$$

$$4x - x > x - x - 8 \qquad \text{Add } -x \text{ to both sides}$$

$$3x > -8$$

$$\frac{1}{3}(3x) > \frac{1}{3}(-8) \qquad \text{Multiply each side by } \frac{1}{3}$$

$$x > -\frac{8}{3}$$

The number is greater than $-\frac{8}{3}$.

65.
$$3x - 5 < x + 7$$

$$3x - 5 + 5 < x + 7 + 5 \qquad \text{Add } 5 \text{ to both sides}$$

$$3x < x + 12$$

$$3x - x < x - x + 12 \qquad \text{Add } -x \text{ to both sides}$$

$$2x < 12$$

$$\frac{1}{2}(2x) < \frac{1}{2}(12) \qquad \text{Multiply each side by } \frac{1}{2}$$

$$x < 6$$

The number is less than 6.

69.
$$\left.\begin{array}{l} x \\ x + 2 \\ x + 4 \end{array}\right\} \quad \text{three sides of the triangle}$$

Remember: To find the perimeter of a triangle, we add all three sides.

$$x + (x + 2) + (x + 4) > 24$$

$$3x + 6 > 24 \qquad \text{Simplify the left side}$$

$$3x + 6 - 6 > 24 - 6 \qquad \text{Add } -6 \text{ to both sides}$$

$$3x > 18$$

$$\frac{1}{3}(3x) > \frac{1}{3}(18) \qquad \text{Multiply each side by } \frac{1}{3}$$

$$x > 6$$

The shortest side is greater than 6 inches.

Section 2.7 continued

73. A (Distributive property)

77. {0, 2} - Remember whole numbers are {0, 1, 2, 3, ...}

81. $\dfrac{130}{858} = \dfrac{2 \cdot 5 \cdot 13}{2 \cdot 3 \cdot 11 \cdot 13} = \dfrac{5}{3 \cdot 11} = \dfrac{5}{33}$

Chapter 2 Review

1. $5x - 8x = (5 - 8)x = -3x$

5. $-a + 2 + 5a - 9$
$= -a + 5a + 2 - 9$
$= (-1 + 5)a + (-7)$
$= 4a - 7$

9. $6 - 2(3y + 1) - 4$
$= 6 - 6y - 2 - 4$
$= -6y$

13. $7x - 2$ letting $x = 3$

   $7(3) - 2 = 19$

17. $-x - 2x - 3x$
$= -6x$ letting $x = 3$
$= -6(3)$
$= -18$

21. $-3x + 2$ letting $x = -2$
$= -3(-2) + 2$
$= 6 + 2$
$= 8$

25. $x + 2 = -6$
$x + 2 + (-2) = -6 + (-2)$
$x = -8$

29. $10 - 3y + 4y = 12$
$10 + y = 12$
$10 + (-10) + y = 12 + (-10)$
$y = 2$

33. $2x = -10$

$\frac{1}{2}(2x) = \frac{1}{2}(-10)$

$x = -5$

37. $\frac{x}{3} = 4$

$3(\frac{x}{3}) = 3(4)$

$x = 12$

41.
$$3a - 2 = 5a$$
$$3a + (-3a) - 2 = 5a + (-3a)$$
$$-2 = 2a$$
$$\frac{1}{2}(-2) = \frac{1}{2}(2a)$$
$$-1 = a$$

45.
$$3x + 2 = 5x - 8$$
$$3x + (-3x) + 2 = 5x + (-3x) - 8$$
$$2 = 2x - 8$$
$$2 + 8 = 2x - 8 + 8$$
$$10 = 2x$$
$$\frac{1}{2}(10) = \frac{1}{2}(2x)$$
$$5 = x$$

49.
$$.7x - .1 = .5x - .1$$
$$.7x + (-.5x) - .1 = .5x + (-.5x) - .1$$
$$.2x - 1 = -.1$$
$$.2x - 1 + 1 = -.1 + 1$$
$$.2x = 0$$
$$\frac{1}{.2}(.2x) = \frac{1}{.2}(0)$$
$$x = 0$$

53.
$$12 = 2(5x - 4)$$
$$12 = 10x - 8$$
$$12 + 8 = 10x - 8 + 8$$
$$20 = 10x$$
$$\frac{1}{10}(20) = \frac{1}{10}(10x)$$
$$2 = x$$

57.
$$\frac{3}{5}(5x - 10) = \frac{2}{3}(9x + 3)$$
$$\frac{3}{5}(5x) - \frac{3}{5}(10) = \frac{2}{3}(9x) + \frac{2}{3}(3)$$
$$3x - 6 = 6x + 2$$
$$3x + (-3x) - 6 = 6x + (-3x) + 2$$
$$-6 = 3x + 2$$
$$-6 + (-2) = 3x + 2 + (-2)$$
$$-8 = 3x$$
$$\left(\frac{1}{3}\right) - 8 = \left(\frac{1}{3}\right)3x$$
$$-\frac{8}{3} = x$$

61. $7 - 3(y + 4) = 10$

$7 - 3y - 12 = 10$

$-3y - 5 = 10$

$-3y - 5 + \mathbf{5} = 10 + \mathbf{5}$

$-3y = 15$

$-\dfrac{1}{3}(-3y) = -\dfrac{1}{3}(15)$

$y = -5$

65. If x is 5, then

$4x - 5y = 20$ becomes

$4(5) - 5y = 20$

$20 - 5y = 20$

$20 + (\mathbf{-20}) - 5y = 20 + (\mathbf{-20})$

$-5y = 0$

$-\dfrac{1}{5}(-5y) = -\dfrac{1}{5}(0)$

$y = 0$

69. $2x - 5y = 10$

$2x + (\mathbf{-2x}) - 5y = \mathbf{-2x} + 10$

$-5y = -2x + 10$

$-\dfrac{1}{5}(-5y) = -\dfrac{1}{5}(-2x + 10)$

$y = -\dfrac{1}{5}(-2x) + -\dfrac{1}{5}(10)$

$y = \dfrac{2x}{5} - 2$

73. What number is 86% of 240?

$x = .86 \cdot 240$

$x = 206.4$

77.

	Now	In 3 years
Bob	$x + 4$	$x + 4 + 3 = x + 7$
Tom	x	$x + 3$

$(x + 7) + (x + 3) = 40$

$2x + 10 = 40$

$2x = 30$

$x = 15$

$x + 4 = 19$

Tom is 15 and Bob is 19 now.

Chapter 2 Review continued

81.

	Dimes	Nickels
number	x	15 - x
value (in cents)	10x	5(15 - x)

(Remember: The total number of coins is 15.)

Amount of money in dimes	+	Amount of money in quarters	=	Total amount of money
10x	+	5(15 - x)	=	100

$$10x + 75 - 5x = 100$$
$$5x + 75 = 100$$
$$5x + 75 + (-75) = 100 + (-75)$$
$$5x = 25$$
$$\frac{1}{5}(5x) = \frac{1}{5}(25)$$

$$x = 5 \qquad \text{dimes}$$
$$15 - x = 10 \qquad \text{nickels}$$

You have 5 dimes and 10 nickels.

85.
$$-2x < 4$$
$$-\frac{1}{2}(-2x) > -\frac{1}{2}(4)$$
$$x > -2$$

89.
$$-\frac{a}{2} \leq -3$$
$$-2(-\frac{a}{2}) \geq -2(-3)$$
$$a \geq 6$$

93.
$$-4x + 5 > -37$$
$$-4x + 5 + (-5) > -37 + (-5)$$
$$-4x > -42$$
$$-\frac{1}{4}(-4x) < -\frac{1}{4}(-42)$$
$$x < \frac{42}{4}$$
$$x < \frac{21}{2}$$

See the graph on page A-7 in the textbook.

97.

$$2x + 10 < 4x - 11$$
$$2x + (-2x) + 10 < 5x + (-2x) - 11$$
$$10 < 3x - 11$$
$$10 + 11 < 3x - 11 + 11$$
$$21 < 3x$$

$$\frac{1}{3}(21) < \frac{1}{3}(3x)$$

$$7 < x \quad \text{or} \quad x > 7$$

See the graph on page A-7 in the textbook.

Chapter 2 Test

1 $3x + 2 - 7x + 3 = 3x - 7x + 2 + 3$ Commutative property
$= (3x - 7x) + (2 + 3)$ Associative property
$= -4x + 5$ Add

2. $4a - 5 - a + 1 = 4a - a - 5 + 1$ Commutative property
$= (4a - a) + (-5 + 1)$ Associative property
$= 3a - 4$

3. $7 - 3(y + 5) - 4 = 7 - 3y - 15 - 4$ Distributive property
$= -3y + 7 - 15 - 4$ Commutative property
$= -3y - 12$ Add

4. $8(2x + 1) - 5(x - 4) = 16x + 8 - 5x + 20$ Distributive property
$= 16x - 5x + 8 + 20$ Commutative property
$= (16x - 5x) + (8 + 20)$ Associative property
$= 11x + 28$ Add

5. $2x - 3 - 7x$ when $x = -5$

$2x - 3 - 7x = 2(-5) - 3 - 7(-5)$
$= -10 - 3 + 35$
$= 22$

6. $x^2 + 2xy + y^2$ when $x = 2$ and $y = 3$

$x^2 + 2xy + y^2 = 2^2 + 2 \cdot 2 \cdot 3 + 3^2$
$= 4 + 2 \cdot 2 \cdot 3 + 9$ Definition of exponents
$= 4 + 12 + 9$ Multiply
$= 25$ Add

7. $2x - 5 = 7$
$2x - 5 + \mathbf{5} = 7 + \mathbf{5}$ Add **5** to both sides
$2x = 12$

$\frac{1}{2}(2x) = \frac{1}{2}(12)$ Multiply each side by $\frac{1}{2}$

$x = 6$

8. $2y + 4 = 5y$
$2y - \mathbf{2y} + 4 = 5y - \mathbf{2y}$ Add **-2y** to both sides
$4 = 3y$

$\frac{1}{3}(4) = \frac{1}{3}(3y)$ Multiply each side by $\frac{1}{3}$

$\frac{4}{3} = y$

Chapter 2 Test continued

9. $\frac{1}{2}x - \frac{1}{10} = \frac{1}{5}x + \frac{1}{2}$

First multiply both sides by LCD which is 10.

$$10\left(\frac{1}{2}x - \frac{1}{10}\right) = 10\left(\frac{1}{5}x + \frac{1}{2}\right)$$

$$10\left(\frac{1}{2}x\right) - 10\left(\frac{1}{10}\right) = 10\left(\frac{1}{5}x\right) + 10\left(\frac{1}{2}\right)$$

$$5x - 1 = 2x + 5$$
$$5x + (\mathbf{-2x}) - 1 = 2x + (\mathbf{-2x}) + 5$$
$$3x - 1 = 5$$
$$3x - 1 + \mathbf{1} = 5 + \mathbf{1}$$
$$3x = 6$$

$$\frac{\mathbf{1}}{\mathbf{3}}(3x) = \frac{\mathbf{1}}{\mathbf{3}}(6)$$

$$x = 2$$

10. $\frac{2}{5}(5x - 10) = -5$

$$\frac{2}{5}(5x) - \frac{2}{5}(10) = -5$$

$$2x - 4 = -5$$
$$2x - 4 + \mathbf{4} = -5 + \mathbf{4}$$
$$2x = -1$$

$$\frac{\mathbf{1}}{\mathbf{2}}(2x) = \frac{\mathbf{1}}{\mathbf{2}}(-1)$$

$$x = -\frac{1}{2}$$

11. $-5(2x + 1) - 6 = 19$ Distributive property
$-10x - 5 - 6 = 19$ Simplify the left side
$-10x - 11 = 19$
$-10x - 11 + \mathbf{11} = 19 + \mathbf{11}$ Add **11** to both sides
$-10x = 30$

$-\frac{\mathbf{1}}{\mathbf{10}}(-10x) = -\frac{\mathbf{1}}{\mathbf{10}}(30)$ Multiply each side by $-\frac{\mathbf{1}}{\mathbf{10}}$

$$x = -3$$

Chapter 2 Test continued

12.
$$.04x + .06(100 - x) = 4.6$$
$$.04x + .06(100) - .06(x) = 4.6$$
$$.04x + 6 - .06x = 4.6$$
$$-.02x + 6 = 4.6$$
$$-.02x + 6 + (\textbf{-6}) = 4.6 + (\textbf{-6})$$
$$-.02x = -1.4$$
$$-\frac{1}{.02}(-.02x) = -\frac{1}{.02}(-1.4)$$
$$x = 70$$

13.
$$2(t - 4) + 3(t + 5) = 2t - 2$$
$$2t - 8 + 3t + 15 = 2t - 2 \qquad \text{Distributive property}$$
$$5t + 7 = 2t - 2 \qquad \text{Simplify the left side}$$
$$5t + 7 - \textbf{7} = 2t - 2 - \textbf{7} \qquad \text{Add } \textbf{-7} \text{ to both sides}$$
$$5t = 2t - 9$$
$$5t - \textbf{2t} = 2t - \textbf{2t} - 9 \qquad \text{Add } \textbf{-2t} \text{ to both sides}$$
$$3t = -9$$
$$\frac{1}{3}(3t) = \frac{1}{3}(-9) \qquad \text{Multiply each side by } \frac{1}{3}$$
$$t = -3$$

14.
$$2x - 4(5x + 1) = 3x + 17$$
$$2x - 20x - 4 = 3x + 17 \qquad \text{Distributive property}$$
$$-18x - 4 = 3x + 17 \qquad \text{Simplify the left side}$$
$$-18x - 4 - \textbf{17} = 3x + 17 - \textbf{17} \qquad \text{Add } \textbf{17} \text{ to both sides}$$
$$-18x - 21 = 3x$$
$$-18x + \textbf{18x} - 21 = 3x + \textbf{18x} \qquad \text{Add } \textbf{18x} \text{ to both sides}$$
$$-21 = 21x$$
$$\frac{1}{21}(-21) = \frac{1}{21}(21x) \qquad \text{Multiply each side by } \frac{1}{21}$$
$$-1 = x$$

15. What number is 15% of 38?
$$x = .15 \quad 38$$
$$x = 5.7$$

16. 240 is 12% of what number?
$$240 = .12x$$
$$\frac{1}{.12}(240) = \frac{1}{.12}(.12x) \qquad \text{Multiply each side by } \frac{1}{.12}$$
$$2,000 = x$$

17.
$$2x - 3y = 12 \qquad y = -2$$
$$2x - 3(-2) = 12$$
$$2x + 6 = 12$$
$$2x + 6 - 6 = 12 - 6 \qquad \text{Add } -6 \text{ to both sides}$$
$$2x = 6$$
$$\frac{1}{2}(2x) = \frac{1}{2}(6) \qquad \text{Multiply each side by } \frac{1}{2}$$
$$x = 3$$

18. $V = \frac{1}{3}\pi r^2 h$ Find h if V = 88 cubic inches, $\pi = \frac{22}{7}$, r = 3 inches

$$88 = \frac{1}{3} \cdot \frac{22}{7} \quad 3^2 h$$
$$88 = \frac{1}{3} \cdot \frac{22}{7} \quad 9h \qquad \text{Definition of exponents}$$
$$88 = \frac{1}{3} \cdot 9 \cdot \frac{22}{7} h \qquad \text{Commutative property}$$
$$88 = 3 \cdot \frac{22}{7} h \qquad \text{Simplify right sides}$$
$$88 = \frac{66}{7} h$$
$$\frac{7}{66}(88) = \frac{7}{66}\left(\frac{66}{7}\right)h \qquad \text{Multiply each side by } \frac{7}{66}$$
$$\frac{7}{3}(4) = h$$
$$\frac{28}{3} = h$$

19.
$$2x + 5y = 20 \text{ for } y$$
$$2x - 2x + 5y = -2x + 20 \qquad \text{Add } -2y \text{ to both sides}$$
$$5y = -2x + 20$$
$$\frac{1}{5}(5y) = \frac{1}{5}(-2x + 20) \qquad \text{Multiply each side by } \frac{1}{5}$$
$$y = -\frac{2}{5}x + 4 \qquad \text{Distributive property}$$

20.
$$h = x + vt + 16t^2 \qquad\qquad \text{for } v$$
$$h - x - 16t^2 = x - x + vt + 16t^2 - 16t^2 \quad \text{Add } -x \text{ and } -16t^2 \text{ to both sides}$$
$$h - x - 16t^2 = vt$$
$$\frac{1}{t}(h - x - 16t^2) = \frac{1}{t}(vt) \qquad\qquad \text{Multiply both sides by } \frac{1}{t}$$
$$\frac{h - x - 16t^2}{t} = v$$

21.

	Age now	Ten years ago
Dave	2x	2x - 10
Rick	x	x - 10

$$(2x - 10) + (x - 10) = 40$$
$$(2x + x) + (-10 - 10) = 40 \qquad \text{Commutative and Associative properties}$$
$$3x - 20 = 40 \qquad \text{Simplify the left side}$$
$$3x - 20 + \mathbf{20} = 40 + \mathbf{20} \qquad \text{Add } \mathbf{20} \text{ to both sides}$$
$$3x = 60$$

$$\frac{1}{3}(3x) = \frac{1}{3}(60) \qquad \text{Multiply each side by } \frac{1}{3}$$

$$x = 20$$

Rick is 20 years old and Dave is 40 years old.

22.
Width   = x
Length   = 2x
Perimeter = 60 inches

$$2 \text{ lengths} + 2 \text{ widths} = \text{perimeter}$$
$$2(2x) + 2x = 60$$
$$4x + 2x = 60 \qquad \text{Multiply}$$
$$6x = 60 \qquad \text{Add}$$

$$\frac{1}{6}(6x) = \frac{1}{6}(60) \qquad \text{Multiply each side by } \frac{1}{6}$$

$$x = 10$$

Width is 10 inches and length is 20 inches.

23.

	dimes	quarters
Number	x + 7	x
Value (in dollars)	.10(x + 7)	.25x

$$.10(x + 7) + .25x = 3.50$$
$$.10x + .70 + .25x = 3.50 \qquad \text{Distributive property}$$
$$.35x + .70 = 3.50 \qquad \text{Commutative and Associative properties}$$
$$.35x + .70 - \mathbf{.70} = 3.50 - \mathbf{.70} \qquad \text{Add } \mathbf{-.70} \text{ to both sides}$$
$$.35x = 2.80$$

$$\frac{1}{.35}(-.35x) = \frac{1}{.35}(2.80) \qquad \text{Multiply each side by } \frac{1}{.35}$$

$$x = 8$$

The man has 8 quarters and 15 dimes.

24.

	Dollars invested at 7%	Dollars invested at 9%
Number of	x	x + 600
Interest on	.07x	.09(x + 600)

$$.07x + .09(x + 600) = 182$$
$$.07x + .09x + 54 = 182 \qquad \text{Distributive property}$$
$$.16x + 54 = 182 \qquad \text{Simplify the left side}$$
$$.16x + 54 - \mathbf{54} = 182 - \mathbf{54} \qquad \text{Add } \mathbf{-54} \text{ to both sides}$$
$$.16x = 128$$

$$\frac{1}{.16}(.16x) = \frac{1}{.16}(128) \qquad \text{Multiply each side by } \frac{1}{.16}$$

$$x = 800$$

The woman invests $800 at 7% and $1,400 at 9%.

25.

$$2x + 3 < 5$$
$$2x + 3 - \mathbf{3} < 5 - \mathbf{3} \qquad \text{Add } \mathbf{-3} \text{ to both sides}$$
$$2x < 2$$

$$\frac{1}{2}(2x) < \frac{1}{2}(2) \qquad \text{Multiply each side by } \frac{1}{2}$$

$$x < 1$$

26.

$$-5a > 20$$
$$-\frac{1}{5}(-5a) > -\frac{1}{5}(20) \qquad \text{Multiply by } -\frac{1}{5}, \text{ and reverse the direction of the inequality symbol}$$
$$a < -4$$

See the graph on page A-7 in the textbook.

27.

$$.4 - .2x \geq 1$$
$$.4 + (-.4) - .2x \geq 1 + (-.4)$$
$$-.2x \geq .6$$
$$-\frac{1}{.2}(-.2x) \geq -\frac{1}{.2}(.6)$$
$$x \leq 3$$

28. $4 - 5(m + 1) \leq 9$

$\qquad 4 - 5m - 5 \leq 9$ Distributive property

$\qquad\qquad -5m - 1 \leq 9$ Commutative and Associative Properties

$\qquad -5m - 1 + 1 \leq 9 + 1$ Add **1** to both sides

$\qquad\qquad\qquad -5m \leq 10$

$\qquad -\dfrac{1}{5}(-5m) \leq -\dfrac{1}{5}(10)$ Multiply by $-\dfrac{1}{5}$ and reverse the direction of the inequality symbol

See the graph on page A-7 in the textbook.

Chapter 3

Section 3.1

1. To complete (0,), substitute x = 0 into 2x + y = 6.
$$2(0) + y = 6$$
$$y = 6$$

The ordered pair is (0, 6).

To complete (3,), substitute x = 3 into 2x + y = 6.
$$2(3) + y = 6$$
$$6 + y = 6$$
$$y = 0$$

The ordered pair is (3, 0).

To complete (, -6), substitute y = -6 into 2x + y = 6.
$$2x - 6 = 6$$
$$2x = 12$$
$$x = 6$$

The ordered pair is (6, -6).

5. To complete (1,), substitute x = 1 into y = 4x - 3.
$$y = 4(1) - 3$$
$$y = 1$$

The ordered pair is (1, 1).

To complete (, 0), substitute y = 0 into y = 4x - 3.
$$0 = 4x - 3$$
$$3 = 4x$$
$$\frac{4}{3} = x$$

The ordered pair is ($\frac{4}{3}$, 0).

To complete (5,), substitute x = 5 into y = 4x - 3.
$$y = 4(5) - 3$$
$$y = 20 - 3$$
$$y = 17$$

The ordered pair is (5, 17).

9. Let x = -5 in each ordered pair.

13. When x = 0, we have
$$y = 4(0)$$
$$y = 0$$

When x = -3, we have
$$y = 4(-3)$$
$$y = -12$$

When y = -2, we have
$$-2 = 4x$$
$$-\frac{1}{2} = x$$

When y = 12, we have
$$12 = 4x$$
$$3 = x$$

See the table on page A-7 of the textbook.

Section 3.1 continued

17.

When y = 0, we have   When y = 2, we have
$$2x - 0 = 4$$ $$2x - 2 = 4$$
$$2x = 4$$ $$2x = 6$$
$$x = 2$$ $$x = 3$$

When $x = \frac{3}{2}$, we have   When x = -3, we have

$$2\left(\frac{3}{2}\right) - y = 4$$ $$2(-3) - y = 4$$
$$3 - y = 4$$ $$-6 - y = 4$$
$$-y = 1$$ $$-y = 10$$
$$y = -1$$ $$y = -10$$

See the table on page A-7 in the textbook.

21.   Try (2, 3) in $2x - 5y = 10$
$$2(2) - 5(3) = 10$$
$$4 - 15 = 10$$
$$-11 = 10$$ A false statement.

Try (0, -2) in $2x - 5y = 10$
$$2(0) - 5(-2) = 10$$
$$0 + 10 = 10$$
$$10 = 10$$ A true statement.

Try $\left(\frac{5}{2}, 1\right)$ in $2x - 5y = 10$

$$2\left(\frac{5}{2}\right) - 5(1) = 10$$

$$5 - 5 = 10$$
$$0 = 10$$ A false statement.

The ordered pair (0, -2) is a solution to the equation

$2x - 5y = 10$; (2, 3) and $\left(\frac{5}{2}, 1\right)$ are not.

25.   Try (1, 6) in $y = 6x$
$$6 = 6(1)$$
$$6 = 6$$ A true statement.
Try (-2, -12) in $y = 6x$
$$-12 = 6(-2)$$
$$-12 = -12$$ A true statement.
Try (0, 0) in $y = 6x$
$$0 = 6(0)$$
$$0 = 0$$ A true statement.

The ordered pairs (1, 6), (-2, -12), (0, 0) are solutions to the equation
$y = 6x$.

Section 3.1 continued

29.   Try $(3, 0)$ in $x = 3$
           $3 = 3$ A true statement.

   Try $(3, -3)$ in $x = 3$
           $3 = 3$ A true statement.

   Try $(5, 3)$ in $x = 3$
           $5 = 3$ A false statement.

   The ordered pairs $(3, 0)$, $(3, -3)$ are solutions to the equation $x = 3$; $(5, 3)$ is not.

33.   When $x = 5$ the ordered pair $(x, 2x)$ becomes $(5, 2 \cdot 5) = (5, 10)$.

37.   When $x = 4$, the equation $3x + 2y = 6$ becomes,

$$3(4) + 2y = 6$$
$$12 + 2y = 6$$
$$2y = -6$$
$$y = -3$$

41.   When $x = 2$, the equation $y = \frac{3}{2}x - 3$ becomes,

$$y = \frac{3}{2}(2) - 3$$
$$y = 3 - 3$$
$$y = 0$$

45. $3x - 2y = 6$
           $-2y = -3x + 6$

$$-\frac{1}{2}(-2y) = -\frac{1}{2}(-3x + 6)$$

$$y = \frac{3}{2}x - 3$$

Section 3.2

For problems 1, 5, 9, 13, and 17 see page A-7 in the textbook.

21. $y = 2x$ $(0, \ )$ $y = 2x$ $(-2, \ )$
           $y = 2(0)$ $y = 2(-2)$
           $y = 0$ $(0, 0)$ $y = -4$ $(-2, -4)$

   $y = 2x$ $(2, \ )$
   $y = 2(2)$ $(2, 4)$

   See the graph on page A-7 in your textbook.

25. $y = 2x + 1$ $\quad$ $(0, \ )$ $\qquad$ $y = 2x + 1$ $\quad$ $(-1, \ )$
$\quad$ $y = 2(0) + 1$ $\qquad\qquad\qquad\quad$ $y = 2(-1) + 1$
$\quad$ $y = 0 + 1$ $\qquad\qquad\qquad\qquad\quad$ $y = -2 + 1$
$\quad$ $y = 1$ $\qquad\quad\ \ $ $(0, 1)$ $\qquad\qquad$ $y = -1$ $\qquad$ $(-1, -1)$

$\quad$ $y = 2x + 1$ $\quad$ $(1, \ )$
$\quad$ $y = 2(1) + 1$
$\quad$ $y = 2 + 1$
$\quad$ $y = 3$ $\qquad\quad\ \ $ $(1, 3)$

See the graph on page A-8 in your textbook.

29. $y = \frac{1}{2}x + 3$ $\qquad$ $(-2, \ )$ $\qquad$ $y = \frac{1}{2}x + 3$ $\qquad$ $(0, \ )$

$\quad$ $y = \frac{1}{2}(-2) + 3$ $\qquad\qquad\qquad\quad$ $y = \frac{1}{2}(0) + 3$

$\quad$ $y = -1 + 3$ $\qquad\qquad\qquad\qquad$ $y = 0 + 3$
$\quad$ $y = 2$ $\qquad\qquad\ $ $(-2, 2)$ $\qquad\qquad$ $y = 3$ $\qquad\qquad$ $(0, 3)$

$\quad$ $y = \frac{1}{2}x + 3$ $\qquad$ $(2, \ )$

$\quad$ $y = \frac{1}{2}(2) + 3$

$\quad$ $y = 1 + 3$
$\quad$ $y = 4$ $\qquad\qquad\ \ $ $(2, 4)$

See the graph on page A-8 in your textbook.

33. $\quad\ $ $2x + y = 3$ $\quad$ $(-1, \ )$ $\qquad$ $2x + y = 3$ $\quad$ $(0, \ )$
$\quad$ $2(-1) + y = 3$ $\qquad\qquad\qquad\ \ $ $2(0) + y = 3$
$\quad\ \ $ $-2 + y = 3$ $\qquad\qquad\qquad\qquad\qquad$ $y = 3$ $\quad$ $(0, 3)$
$\qquad\qquad\ $ $y = 5$ $\quad$ $(-1, 5)$

$\quad\ \ $ $2x + y = 3$ $\quad$ $(1, \ )$
$\quad\ $ $2(1) + y = 3$
$\quad\quad\ $ $2 + y = 3$
$\qquad\qquad\ $ $y = 1$ $\quad$ $(1, 1)$

See the graph on page A-8 in your textbook.

37. $\quad\ $ $-x + 2y = 6$ $\quad$ $(-2, \ )$ $\qquad$ $-x + 2y = 6$ $\quad$ $(0, \ )$
$\quad$ $-(-2) + 2y = 6$ $\qquad\qquad\qquad\ \ $ $-0 + 2y = 6$
$\quad\quad\ \ $ $2 + 2y = 6$ $\qquad\qquad\qquad\qquad\ \ $ $2y = 6$
$\qquad\qquad\ \ $ $2y = 4$ $\qquad\qquad\qquad\qquad\ \ $ $y = 3$ $\quad$ $(0, 3)$
$\qquad\qquad\qquad$ $y = 2$ $\quad$ $(-2, 2)$

$\quad\quad\ $ $-x + 2y = 6$ $\quad$ $(2, \ )$
$\quad\quad\ \ $ $-2 + 2y = 6$
$\qquad\qquad\ \ $ $2y = 8$
$\qquad\qquad\qquad$ $y = 4$ $\quad$ $(2, 4)$

See the graph on page A-8 in your textbook.

41. To find three solutions to $y = 3x - 1$ we can let $x = 1$, $x = 0$, and $x = -1$, and find the corresponding values of y.

When $x = 1$, $y = 3(1) - 1 = 2$. The ordered pair $(1, 2)$ is one solution.

When $x = 0$, $y = 3(0) - 1 = -1$. The ordered pair $(0, -1)$ is a second solution.

When $x = -1$, $y = 3(-1) - 1 = -4$. The ordered pair $(-1, -4)$ is a third solution.

The graph is on page A-9 in the textbook.

45. To find three solutions to $3x + 4y = 8$ we can let $x = 0$, $x = 1$, and $y = 0$.

When $x = 0$, $\quad 3(0) + 4y = 8$
$\qquad\qquad\qquad 4y = 8$
$\qquad\qquad\qquad\ y = 2 \qquad$ so $(0, 2)$ is one solution.

When $x = 1$, $\quad 3(1) + 4y = 8$
$\qquad\qquad\quad 3 + 4y = 8$
$\qquad\qquad\qquad 4y = 5$

$\qquad\qquad\qquad\ y = \dfrac{5}{4} \qquad$ so $(1, \dfrac{5}{4})$ is a second solution.

When $y = 0$, $\quad 3x + 4(0) = 8$
$\qquad\qquad\qquad 3x = 8$

$\qquad\qquad\qquad\ x = \dfrac{8}{3} \qquad$ so $(\dfrac{8}{3}, 0)$ is a third solution.

The graph is on page A-9 in your textbook.

49. $y = 2 \qquad x$ may equal any real number.

See the graph on page A-9 in your textbook.

53. Some suggested points

$\qquad (0, 0)$, $(1, 1)$, $(-1, -1)$, $(2, 2)$

See the graph on page A-9 in your textbook.

57. The lines cross the y-axis at the number that follows x in the equations.

See the graph on page A-9 in your textbook.

61. $2(3x - 1) + 4 = -10$
$\qquad 6x - 2 + 4 = -10 \qquad$ Distributive property
$\qquad\quad 6x + 2 = -10$
$\qquad\qquad\quad 6x = -12 \qquad$ Add -2 to both sides

$\qquad\qquad\qquad x = -2 \qquad$ Multiply each side by $\dfrac{1}{6}$

Section 3.2 continued

65. $\frac{1}{2}x + 4 = \frac{2}{3}x + 5$

Method 1: Working with the fractions.

$$\frac{1}{2}x + (-\frac{1}{2}x) + 4 = \frac{2}{3}x + (-\frac{1}{2}x) + 5$$

$$4 = \frac{1}{6}x + 5 \qquad\qquad \frac{2}{3} - \frac{1}{2} = \frac{4}{6} - \frac{3}{6} = \frac{1}{6}$$

$$-1 = \frac{1}{6}x \qquad\qquad \text{Add } -5 \text{ to both sides}$$

$$-6 = x \qquad\qquad \text{Multiply each side by 6}$$

Method 2: Eliminating the fractions in the beginning.

$$6(\frac{1}{2}x + 4) = 6(\frac{2}{3}x + 5) \qquad \text{Multiply each side by the LCD 6}$$

$$6(\frac{1}{2}x) + 6(4) = 6(\frac{2}{3}x) + 6(5) \qquad \text{Distributive property}$$

$$3x + 24 = 4x + 30 \qquad \text{Multiply}$$
$$24 - x + 30 \qquad\qquad \text{Add } -3x \text{ to each side}$$
$$-6 = x \qquad\qquad \text{Add } -30 \text{ to each side}$$

Section 3.3

1. x-intercept

When $\qquad\qquad\qquad y = 0$
the equation $\qquad 2x + y = 4$
becomes $\qquad\quad 2x + 0 = 4$
$\qquad\qquad\qquad\qquad 2x = 4$

$\qquad\qquad\qquad\qquad\ \ x = 2 \qquad$ Multiply each side by $\frac{1}{2}$

The x-intercept is 2, so the point (2, 0) is on the graph.

y-intercept

When $\qquad\qquad\qquad x = 0$
the equation $\qquad 2x + y = 4$
becomes $\qquad\quad 2(0) + y = 4$
$\qquad\qquad\qquad\qquad\ y = 4$

The y-intercept is 4, so the point (0, 4) is on the graph.

See the graph on page A-10 in your textbook.

Section 3.3 continued

5. **x-intecept**

    When $y = 0$
    the equation $-x + 2y = 2$
    becomes $-x + 2(0) = 2$
    $-x = 2$
    $x = -2$

    The x-intercept is -2, so the point (-2, 0) is on the graph.

    y-intercept

    When $x = 0$
    the equation $-x + 2y = 2$
    becomes $-(0) + 2y = 2$
    $y = 1$

    The y-intercept is 1, so the point (0, 1) is on the graph.

    See the graph on page A-10 in your textbook.

9. **x-intercept**

    When $y = 0$
    the equation $4x - 2y = 8$
    becomes $4x - 2(0) = 8$
    $4x = 8$

    $x = 2$ Multiply each side by $\frac{1}{2}$

    The x-intercept is 2, so the point (2, 0) is on the graph.

    y-intercept

    When $x = 0$
    the equation $4x - 2y = 8$
    becomes $4(0) - 2y = 8$
    $-2y = 8$
    $y = -4$ Multiply each side by $-\frac{1}{2}$

    The y-intercept is -4, so the point (0, -4) is on the graph.

    See the graph on page A-10 in your textbook.

13. **x-intercept**

    When $y = 0$
    the equation $y = 2x - 6$
    becomes $0 = 2x - 6$
    $6 = 2x$
    $3 = x$ Multiply each side by $\frac{1}{2}$

    The x-intercept is 3, so the point (3, 0) is on the graph.

    y-intercept

    When $x = 0$
    the equation $y = 2x - 6$
    becomes $y = 2(0) - 6$
    $y = -6$

    The y-intercept is -6, so the point (0, -6) is on the graph.

    See the graph on page A-10 in your textbook.

17. **x-intercept**

When $\quad y = 0$

the equation $\quad y = 2x - 1$

becomes $\quad 0 = 2x - 1$

$$1 = 2x$$

$$\frac{1}{2} = x \qquad \text{Multiply each side by } \frac{1}{2}$$

The x-intercept is $\frac{1}{2}$, so the point ($\frac{1}{2}$, 0) is on the graph.

y-intercept

When $\quad x = 0$

the equation $\quad y = 2x - 1$

becomes $\quad y = 2(0) - 1$

$$y = -1$$

The y-intercept is -1, so the point (0, -1) is on the graph.

See the graph on A-10 in your textbook.

21. **x-intercept**

When $\quad y = 0$

the equation $\quad y = -\frac{1}{3}x - 2$

becomes $\quad 0 = -\frac{1}{3}x - 2$

$$2 = -\frac{1}{3}x$$

$$-6 = x \qquad \text{Multiply each side by -3}$$

The x-intercept is -6, so the point (-6, 0) is on the graph.

y-intercept

When $\quad x = 0$

the equation $\quad y = -\frac{1}{3}x - 2$

becomes $\quad y = -\frac{1}{3}(0) - 2$

$$y = -2$$

The y-intercept is -2, so the point (0, -2) is on the graph.

See the graph on page A-11 in your textbook.

25. See the graph on page A-11 in the textbook.

29. See the graph on page A-11 in the textbook.

33. The y-intercept is -4. See the graph on page A-11 in the textbook.

Section 3.3 continued

37. The x-intercept is 3 and the y-intercept is 3. See the graph on page A-12 in the textbook.

41. The y-intercept is 4.

See the graph on page A-12 in your textbook.

45. $5 - 2 \cdot 6 = 5 - 12$ Multiply

           $= -7$ Subtract

49. $2(3)^2 - 4(3)^2 = 2(9) - 4(9)$ Definition of exponents

                  $= 18 - 36$ Multiply

                  $= -18$ Subtract

Section 3.4

1. Let $(x_1, y_1) = (2,1)$ and $(x_2, y_2) = (4,4)$.

$$m = \frac{y_2 - y_1}{x_2 - x_1} = \frac{4 - 1}{4 - 2} = \frac{3}{2}$$

See the graph on page A-12 in the textbook.

5. Let $(x_1, y_1) = (1,-3)$ and $(x_2, y_2) = (4,2)$.

$$m = \frac{y_2 - y_1}{x_2 - x_1} = \frac{2 - (-3)}{4 - 1} = \frac{5}{3}$$

See the graph on page A-12 in the textbook.

9. Let $(x_1, y_1) = (-3,2)$ and $(x_2, y_2) = (3,-2)$.

$$m = \frac{y_2 - y_1}{x_2 - x_1} = \frac{-2 - 2}{3 - (-3)} = \frac{-4}{6} = -\frac{2}{3}$$

See the graph on page A-12 in the textbook.

13. Slope $= \frac{2}{3}$ y-intercept $= 1$

See the graph on page A-13 in your textbook.

17. Slope $= -\frac{4}{3}$ y-intercept $= 5$

See the graph on page A-13 in the textbook.

21. Slope $= -1$ y-intercept $= 3$

See the graph on page A-13 in the textbook.

25. The x-intercept is 4, so the point (4, 0) is on the graph.
The y-intercept is 2, so the point (0, 2) is on the graph.

$$m = \frac{y_2 - y_1}{x_2 - x_1} = \frac{2 - 0}{0 - 4} = \frac{2}{-4} = -\frac{2}{4} = -\frac{1}{2}$$

See the graph on page A-13 in your textbook.

29. $y = mx + b$

$$y = \frac{1}{2}x + 1$$

$m = \frac{1}{2}$ (slope) $b = 1$ (y-intercept)

See the graph on page A-13 in the textbook.

33. $V = IR$

$\dfrac{V}{R} = \dfrac{IR}{R}$ Divide both sides by R

$\dfrac{V}{R} = I$

37.
$$-2x + y = -4$$
$$-2x + 2x + y = 2x - 4 \qquad \text{Add } 2x \text{ to both sides}$$
$$y = 2x - 4$$

Section 3.5

1. Slope = $\frac{2}{3}$ y-intercept = 1

See the graph on page A-14 in your textbook.

5. Slope = $-\frac{2}{5}$ y-intercept = 3

See the graph on page A-14 in your textbook.

9. Slope = $-\frac{3}{1}$ y-intercept = 2

See the graph on page A-14 in your textbook.

13.
$$3x + y = 3 \qquad \text{Original equation}$$
$$\mathbf{3x} - 3x + y = \mathbf{3x} + 3 \qquad \text{Add } \mathbf{3x} \text{ to both sides}$$
$$y = 3x + 3 \qquad \text{Equation solved for } y. \text{ Now in slope intercept form.}$$

Slope = $\frac{3}{1}$ y-intercept = 3

See the graph on page A-14 in your textbook.

Section 3.5 continued

17.
$$4x - 5y = 20 \qquad \text{Original equation}$$
$$4x - \mathbf{4x} - 5y = -\mathbf{4x} + 20 \qquad \text{Add } \mathbf{-4x} \text{ to both sides}$$
$$-5y = -4x + 20$$

$$-\frac{1}{5}(-5y) = -\frac{1}{5}(-4x + 20) \qquad \text{Multiply each side by } -\frac{1}{5}$$

$$y = \frac{4}{5}x - 4$$

$$\text{Slope} = \frac{4}{5} \qquad \text{y-intercept} = -4$$

See the graph on page A-14 in your textbook.

21.   When $\qquad x = -2, \ y = -5, \ m = 2$
The equation $\qquad y = mx + b$
becomes $\qquad -5 = 2(-2) + b$
$\qquad\qquad\qquad -5 = -4 + b \qquad\qquad$ Multiply
$\qquad\qquad\qquad -1 = b \qquad\qquad\qquad$ Add 4 to both sides

The complete equation is $y = 2x - 1$ because the slope is 2, and the y-intercept is -1.

25.   When $\qquad x = 2, \ y = -3, \ m = \frac{3}{2}$

the equation $\qquad y = mx + b$

becomes $\qquad -3 = \frac{3}{2}(2) + b$

$\qquad\qquad\qquad -3 = 3 + b \qquad\qquad$ Multiply
$\qquad\qquad\qquad -6 = b \qquad\qquad\qquad$ Add +1 to both sides

The complete equation is $y = \frac{3}{2}x - 6$ because the slope is $\frac{3}{2}$, and the y-intercept is -6.

29.   When $\qquad x = 2, \ y = 4, \ m = 1$
the equation $\qquad y = mx + b$
becomes $\qquad 4 = 1(2) + b$
$\qquad\qquad\qquad 4 = 2 + b \qquad\qquad$ Multiply
$\qquad\qquad\qquad 2 = b \qquad\qquad\qquad$ Add -2 to both sides

The complete equation is $y = 1x + 2$ or $y = x + 2$ because the slope is 1 and the y-intercept is 2.

33. $m = \dfrac{y_2 - y_1}{x_2 - x_1} = \dfrac{1 - (-5)}{2 - (-1)} = \dfrac{6}{3} = 2$

When $\qquad x = -1, \ y = -5, \ m = 2$
the equation $\qquad y = mx + b$
becomes $\qquad -5 = 2(-1) + b$
$\qquad\qquad\qquad -5 = -2 + b \qquad\qquad$ Multiply
$\qquad\qquad\qquad -3 = b \qquad\qquad\qquad$ Add 2 to both sides

The completed equation is $y = 2x - 3$ because the slope is 2 and the y-intercept is -3.

37. $m = \dfrac{y_2 - y_1}{x_2 \quad x_1} = \dfrac{-5 - (-1)}{3 - (-3)} = \dfrac{-4}{6} = -\dfrac{2}{3}$

When $\qquad x = -3, \ y = -1, \ m = -\dfrac{2}{}$ $\qquad$ (We could use x = 3 and
$\qquad\qquad\qquad\qquad\qquad\qquad\qquad\qquad\qquad\qquad\qquad$ y = -5.)

the equation $\qquad y = mx + b$

becomes $\qquad\qquad -1 = -\dfrac{2}{3}(-3) + b$

$\qquad\qquad\qquad\qquad -1 = 2 + b \qquad\qquad$ Multiply
$\qquad\qquad\qquad\qquad -3 = b \qquad\qquad\quad$ Add -1 to both sides

The completed equation is $y = -\dfrac{2}{3}x - 3$, because the slope $-\dfrac{2}{3}$ and
the y-intercept is -3.

41. The x-intercept is 3 so the point (3, 0) is on the graph.
The y-intercept is 2 so the point (0, 2) is on the graph.

$\qquad m = \dfrac{2 - 0}{0 - 3} = \dfrac{2}{-3} = -\dfrac{2}{3}$

$\qquad y = mx + b$

$\qquad y = -\dfrac{2}{3}x + 2$

45. x = 3, because x has the value of 3 in the given points (3, 2) and
(3, 4).

49. What number is 25% of 300?

$\qquad x = .25 \cdot 300$
$\qquad x = .25(300)$
$\qquad x = 75$

53. 60 is 15% of what number?

$\qquad 60 = .15 \cdot x$
$\qquad 60 = .15x$
$\qquad 400 = x \qquad\qquad$ Multiply each side by $\dfrac{1}{.15}$

57. To graph y = 2x - 3, we find its intercepts.

x-intercept
$\qquad$ When $\qquad\qquad y = 0$
$\qquad$ the equation $\quad y = 2x - 3$
$\qquad$ becomes $\qquad\quad 0 = 2x - 3$
$\qquad\qquad\qquad\qquad\quad 3 = 2x$

$\qquad\qquad\qquad\qquad\quad \dfrac{3}{2} = x$

$\qquad\qquad\qquad\qquad (\dfrac{3}{2}, 0)$

y-intercept
$\qquad$ When $\qquad\qquad x = 0$
$\qquad$ the equation $\quad y = 2x - 3$
$\qquad$ becomes $\qquad\quad y = 2(0) \ 3$
$\qquad\qquad\qquad\qquad\quad y = -3$
$\qquad\qquad\qquad\qquad (0, -3)$

See the graph on page A-14 in the textbook.

Section 3.6

1. To graph x + y = 3 we find its intercepts.

   x-intercept   **y-intercept**
   When   y = 0   When   x = 0
   the equation   x + y = 3   the equation   x + y = 3
   becomes   x + 0 = 3   becomes   0 + y = 3
                      x = 3   y = 3
                    (3, 0)   (0, 3)

   To graph x - y = 1 we find its intercepts.

   x-intercept   **y-intercept**
   When   y = 0   When   x = 0
   the equation   x - y = 1   the equation   x - y = 1
   becomes   x - 0 = 1   becomes   0 - y = 1
                      x = 1   y = -1
                    (1, 0)   (0, -1)

   The solution to the system is the point (2, 1).

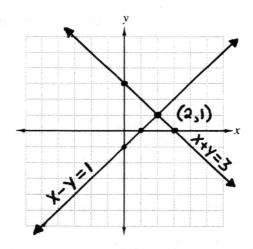

5.   To graph x + y = 8 we find its intercepts.

   x-intercept   **y-intercept**
   When   y = 0   When   x = 0
   the equation   x + y = 8   the equation   x + y = 8
   becomes   x + 0 = 8   becomes   0 + y = 8
                          x = 8   y = 8
                        (8, 0)   (0, 8)

   To graph y = x + 2, we find its intercepts.

   x-intercept   **y-intercept**
   When   y = 0   When   x = 0
   the equation   y = x + 2   the equation   y = x + 2
   becomes   0 = x + 2   becomes   y = 0 + 2
                        -2 = x   y = 2
                       (-2,0)   (0,2)

   The solution to the system is the point (3,5).

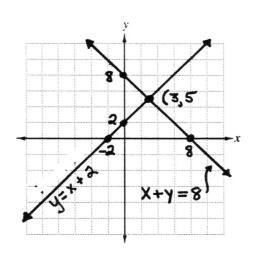

9. To graph 6x - 2y = 12 we find its intercepts.

x-intercept

When $y = 0$
the equation $6x - 2y = 12$
becomes $6x - 2(0) = 12$
$6x = 12$
$x = 2$
$(2, 0)$

y-intercept

When $x = 0$
the equation $6x\ 2y = 12$
becomes $6(0)\ 2y = 12$
$-2y = 12$
$y = -6$
$(0, -6)$

To graph 3x + y = -6 we find its intercepts.

x-intercept

When $y = 0$
the equation $3x + y = -6$
becomes $3x + 0 = -6$
$3x = -6$
$x = -2$
$(-2, 0)$

y-intercept

When $x = 0$
the equation $3x + y = -6$
becomes $3(0) + y = -6$
$y = -6$
$(0, -6)$

The solution to the system is the point (0, -6).

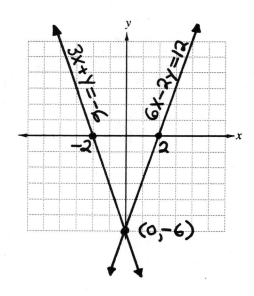

13.   To graph x + 2y = 0 we find the x-intercept and another point.

    x-intercept

When	$y = 0$	When	$y = 1$
the equation	$x + 2y = 0$	the equation	$x + 2y = 0$
becomes	$x + 2(0) = 0$	becomes	$x + 2(1) = 0$
	$x = 0$		$x + 2 = 0$
	$(0,0)$		$x = -2$

so (-2, 1) is a point on the graph.

To graph y = 2x, we find the x-intercept and another point.

    x-intercept

When	$y = 0$	When	$x = 1$
the equation	$y = 2x$	the equation	$y = 2x$
becomes	$0 = 2x$	becomes	$y = 2(1)$
	$0 = x$		$y = 2$
	$(0,0)$		

so (1,2) is a point on the graph.

The solution to the system is the point (0,0).

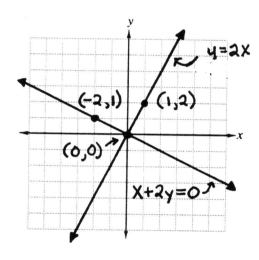

17. To graph $y = 2x + 1$ we find the intercepts.

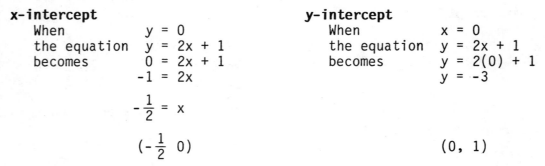

x-intercept

When $y = 0$
the equation $y = 2x + 1$
becomes $0 = 2x + 1$

$-1 = 2x$

$-\dfrac{1}{2} = x$

$\left(-\dfrac{1}{2}\ 0\right)$

y-intercept

When $x = 0$
the equation $y = 2x + 1$
becomes $y = 2(0) + 1$

$y = -3$

$(0,\ 1)$

To graph $y = -2x - 3$ we find the intercepts.

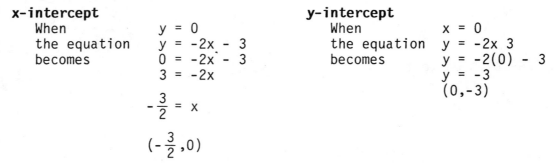

x-intercept

When $y = 0$
the equation $y = -2x - 3$
becomes $0 = -2x - 3$

$3 = -2x$

$-\dfrac{3}{2} = x$

$\left(-\dfrac{3}{2}, 0\right)$

y-intercept

When $x = 0$
the equation $y = -2x\ 3$
becomes $y = -2(0) - 3$

$y = -3$

$(0, -3)$

The solution to the system is the point $(-1, -1)$.

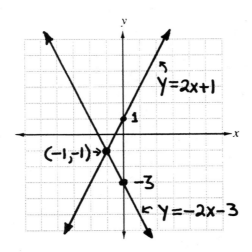

21. For x + y = 2, the intercepts are:

x-intercept

When   y = 0
the equation   x + y = 2
becomes   x + 0 = 2
                         x = 2

y-intercept

When   x = 0
the equation   x + y = 2
becomes   0 + y = 2
                         y = 2

x = -3 is a vertical line. y is all real numbers.
The solution to the system is the point (-3, 5).

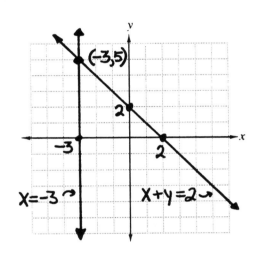

Section 3.6 continued

25. For x + y = 4, the intercepts are:

x-intercept   **y-intercept**
    When   y = 0   When   x = 0
    the equation   x + y = 4   the equation   x + y = 4
    becomes   x + 0 = 4   becomes   0 + y = 4
                          x = 4   y = 4
                        (4,0)   (0,4)

For 2x + 2y = -6, the intercepts are:

x-intercept   **y-intercept**
    When   y = 0   When   x = 0
    the equation   2x + 2y = -6   the equation   2x + 2y = -6
    becomes   2x + 2(0) = -6   becomes   2(0) + 2y = -6
                           x = -3   y = -3
                        (-3,0)   (0,-3)

Lines are parallel. There is no solution to the system.

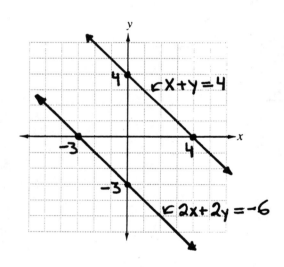

Section 3.6 continued

29. Substituting $x = -3$ into $2x - 9$ we have
$$2(-3) - 9 = -6 - 9 = -15$$

33. Substituting $x = -3$ into $2x^2 + 3x + 4$, we have
$$2(-3)^2 + 3(-3) + 4 = 2(9) + 3(-3) + 4$$
$$= 18 - 9 + 4$$
$$= 9 + 4$$
$$= 13$$

37. $8 - 2(4 - 3x) = 8 - 8 + 6x$　　　Distributive property
$$= 6x$$

Section 3.7

1.　$\begin{aligned} x + y &= 3 \\ \underline{x - y} &= \underline{1} \\ 2x &= 4 \end{aligned}$　　Add together

　　　　$x = 2$　　Divide by 2

Substituting $x = 2$ into $x + y = 3$ gives us
$$2 + y = 3$$
$$y = 1$$

The solution to our system is the ordered pair (2, 1).

5.　$\begin{aligned} x - y &= 7 \\ \underline{-x - y} &= \underline{3} \\ -2y &= 10 \end{aligned}$　　Add together

　　　　$y = -5$　　Divide by -2

Substituting $y = -5$ into $x - y = 7$ gives us
$$x - (-5) = 7$$
$$x + 5 = 7$$
$$x = 2$$

The solution to our system is the ordered pair (2, -5).

9.　$\begin{aligned} 3x + 2y &= 1 \\ \underline{-3x + -2y} &= \underline{1} \\ 0 &= 0 \end{aligned}$　　Add together

The lines coincide and there is an infinite number of solutions to the system.

13. $\begin{aligned} 5x - 3y &= -2 \\ 10x - y &= 1 \end{aligned}$ $\xrightarrow[\text{no change}]{\text{multiply by -2}}$ $\begin{aligned} -10x + 6y &= 4 \\ \underline{10x - y = 1} \\ 5y &= 5 \\ y &= 1 \end{aligned}$

Substituting $y = 1$ into $10x - y = 1$ gives us

$$10x - 1 = 1$$
$$10x = 2$$
$$x = \frac{2}{10}$$
$$x = \frac{1}{5}$$

The solution to our system is the ordered pair $(\frac{1}{5}, 1)$.

17. $\begin{aligned} 3x - 5y &= 7 \\ -x + y &= -1 \end{aligned}$ $\xrightarrow[\text{multiply by 3}]{\text{no change}}$ $\begin{aligned} 3x - 5y &= 7 \\ \underline{-3x + 3y = -3} \\ -2y &= 4 \\ y &= -2 \end{aligned}$

Substituting $y = -2$ into $-x + y = -1$ gives us

$$-x - 2 = -1$$
$$-x = 1$$
$$x = -1$$

The solution to our system is the ordered pair $(-1, -2)$.

21. $\begin{aligned} -3x - y &= 7 \\ 6x + 7y &= 11 \end{aligned}$ $\xrightarrow[\text{no change}]{\text{multiply by 2}}$ $\begin{aligned} -6x - 2y &= 14 \\ \underline{6x + 7y = 11} \\ 5y &= 25 \\ y &= 5 \end{aligned}$

Substituting $y = 5$ into $-3x - y = 7$ gives us

$$-3x - 5 = 7$$
$$-3x = 12$$
$$x = -4$$

The solution to our system is the ordered pair $(-4, 5)$.

25. $\begin{aligned} x + 3y &= 9 \\ 2x - y &= 4 \end{aligned}$ $\xrightarrow[\text{multiply by 3}]{\text{no change}}$ $\begin{aligned} x + 3y &= 9 \\ \underline{6x - 3y = 12} \\ 7x &= 21 \\ x &= 3 \end{aligned}$

Substituting $x = 3$ into $x + 3y = 9$ gives us

$$3 + 3y = 9$$
$$3y = 6$$
$$y = 2$$

The solution to our system is the ordered pair $(3, 2)$.

29.

$$
\begin{array}{l}
2x + 9y = 2 \xrightarrow{\text{no change}} \quad 2x + 9y = 2 \\
5x + 3y = -8 \xrightarrow{\text{multiply by -3}} -15x - 9y = 24 \\
\hline
\qquad\qquad\qquad\qquad\qquad -13x = 26 \\
\qquad\qquad\qquad\qquad\qquad\quad x = -2
\end{array}
$$

Substituting $x = -2$ into $2x + 9y = 2$ gives us

$$
\begin{aligned}
2(-2) + 9y &= 2 \\
-4 + 9y &= 2 \\
9y &= 6 \\
y &= \frac{6}{9} \\
y &= \frac{2}{3}
\end{aligned}
$$

The solution to our system is the ordered pair $(-2, \frac{2}{3})$.

33.

$$
\begin{array}{l}
3x + 2y = -1 \xrightarrow{\text{multiply by -2}} -6x + -4y = 25 \\
6x + 4y = 0 \xrightarrow{\text{no change}} \quad 6x + 4y = 0 \\
\hline
\qquad\qquad\qquad\qquad\qquad\qquad\quad 0 = 2
\end{array}
$$

The lines are parallel and there is no solution to the system.

37. $\frac{1}{2}x + \frac{1}{6}y = \frac{1}{3} \xrightarrow{\text{multiply by 6}} 3x + y = 2$

$-x - \frac{1}{3}y = -1 \xrightarrow{\text{multiply by 6}} -6x - 2y = -1$

$$
\begin{array}{l}
3x + y = 2 \xrightarrow{\text{multiply by 2}} \quad 6x + 2y = 4 \\
-6x - 2y = -1 \xrightarrow{\text{no change}} -6x - 2y = -1 \\
\hline
\qquad\qquad\qquad\qquad\qquad\qquad\quad 0 = 3
\end{array}
$$

The lines are parallel and there is no solution to the system.

41.

$$
\begin{array}{l}
x + y = 22 \xrightarrow{\text{no change}} \quad x + y = 22 \\
0.05x + 0.10y = 1.70 \xrightarrow{\text{multiply by 100}} 5x + 10y = 170
\end{array}
$$

$$
\begin{array}{l}
x + y = 22 \xrightarrow{\text{multiply by -5}} -5x - 5y = -110 \\
5x + 10y = 17 \xrightarrow{\text{no change}} \quad 5x + 10y = 170 \\
\hline
\qquad\qquad\qquad\qquad\qquad\qquad 5y = 60 \\
\qquad\qquad\qquad\qquad\qquad\qquad\; y = 12
\end{array}
$$

Substituting $y = 12$ into $x + y = 22$ gives us

$$
\begin{aligned}
x + 12 &= 22 \\
x &= 10
\end{aligned}
$$

The solution to our system is the ordered pair $(10, 12)$.

45.

$$-3x \geq 12$$

$$-\frac{1}{3}(-3x) \leq -\frac{1}{3}(12)$$

Multiply each side by $-\frac{1}{3}$ and reverse the direction of the inequality symbol.

$$x \leq -4$$

49.

$$-4x + 1 < 17$$
$$-4x < 16$$

Add -1 to both sides

$$-\frac{1}{4}(-4x) > -\frac{1}{4}(16)$$

Multiply each side by $-\frac{1}{4}$ and reverse the direction of the inequality symbol.

$$x > -4$$

Section 3.8

1. $x + y = 11$
$\quad\quad y = 2x - 1$

Substituting the expression $2x - 1$ for y from the second equation into the first equation, we have

$$x + (2x - 1) = 11$$
$$3x - 1 = 11$$
$$3x = 12$$
$$x = 4$$

When $x = 4$, $\quad y = 2x - 1$
becomes $\quad\quad y = 2(4) - 1 = 8 - 1 = 7$

The solution to our system is (4, 7).

5. $-2x + y = -1$
$\quad\quad y = -4x + 8$

Substituting the expression $-4x + 8$ for y from the second equation into the first equation, we have

$$-2x + (-4x + 8) = -1$$
$$-6x + 8 = -1$$
$$-6x = -9$$
$$x = \frac{-9}{-6}$$
$$x = \frac{3}{2}$$

When $x = -\frac{3}{2}$, $\quad\quad\quad y = -4x + 8$

becomes $\quad\quad\quad -4(\frac{3}{2}) + 8 = -6 + 8 = 2$

The solution to our system is $(\frac{3}{2}, 2)$.

9. $5x - 4y = -16$
    $y = 4$

Substituting the expression 4 for y from the second equation into the first equation, we have

$$5x - 4(4) = -16$$
$$5x - 16 = -16$$
$$5x = 0$$
$$x = 0$$

When $x = 0$, $y = 4$, the solution to our system is $(0, 4)$.

13. $x + 3y = 4$
    $x - 2y = -1$

Solving the second equation for x gives us $x = 2y - 1$.

Substituting the expression $2y - 1$ for x into the first equation, we have

$$(2y - 1) + 3y = 4$$
$$5y - 1 = 4$$
$$5y = 5$$
$$y = 1$$

When $y = 1$, $x = 2y - 1$
becomes $x = 2(1) - 1 = 2 - 1 = 18 = 8$

The solution to our system is $(1, 1)$.

17. $3x + 5y = -3$
    $x - 5y = -5$

Solving the second equation for x gives us $x = 5y - 5$

Substituting the expression $5y - 5$ for x into the first equation, we have

$$3(5y - 5) + 5y = -3$$
$$15y - 15 + 5y = -3$$
$$20y - 15 = -3$$
$$20y = 12$$
$$y = \frac{12}{20}$$
$$y = \frac{3}{5}$$

When $y = \frac{3}{5}$, $x - 5y = -5$

becomes $x - 5(\frac{3}{5}) = -5$

$$x - 3 = -5$$
$$x = -2$$

The solution to our system is $(-2, \frac{3}{5})$.

Section 3.8 continued

21. $-3x - 9y = 7$
    $x + 3y = 2$

Solving the second equation for x gives us $x = -3y + 2$.

Substituting the expression $-3y + 2$ for x into the first equation, we have

$$-3(-3y + 2) - 9y = 7$$
$$9y - 6 - 9y = 7$$
$$-6 = 7 \qquad \text{False}$$

The lines are parallel and there is no solution to the system.

25. $7x - 6y = -1$
    $x = 2y - 1$

Substituting the expression $2y - 1$ for x from the second equation into the first equation, we have

$$7(2y - 1) - 6y = -1$$
$$14y - 7 - 6y = -1$$
$$8y - 7 = -1$$
$$8y = 6$$
$$y = \frac{6}{8}$$
$$y = \frac{3}{4}$$

When $y = \frac{3}{4}$, $x = 2y - 1$

becomes $x = 2(\frac{3}{4}) - 1 = \frac{3}{2} - 1 = \frac{1}{2}$

The solution to our system is $(\frac{1}{2}, \frac{3}{4})$

29. $5x - 6y = -4$
    $x = y$

Substituting the expression y for x from the second equation into the first equation, we have

$$5y - 6y = -4$$
$$-y = -4$$
$$y = 4$$

Since x and y are equal, x must be 4 also.

The solution to our system is (4, 4).

33. $7x - 11y = 16$
$\qquad\qquad y = 10$

Substituting the expression 10 for y from the second equation into the first equation, we have

$$7x - 11(10) = 16$$
$$7x - 110 = 16$$
$$7x = 126$$
$$x = 18$$

The solution to our system is (18, 10).

37. Substituting $y = 22 - x$ into the first equation, we have

$$0.05x + 0.10(22 - x) = 1.70 \qquad \text{Distributive property}$$
$$0.05x + 2.2 - 0.10x = 1.70$$
$$2.2 - 0.05x = 1.70$$
$$-0.05x = -0.50 \qquad \text{Add -2.2 to both sides}$$
$$x = 10 \qquad \text{Divide both sides by -0.05}$$

By first eliminating the decimals, we have

$$5x + 10y = 170$$

Substituting $y = 22 - x$ into this equation, we have

$$5x + 10(22 - x) = 170$$
$$5x + 220 - 10x = 170 \qquad \text{Distributive property}$$
$$220 - 5x = 170$$
$$-5x = -50 \qquad \text{Add -220 to both sides}$$
$$x = 10 \qquad \text{Divide both sides by -5}$$
$$y = 22 - 10$$
$$y = 12$$

The solution is (10,12).

41.

	Nickles	Dimes
Number of	x	x + 3
Value of	.05x	.10(x + 3)

$$.05x + .10(x + 3) = 2.10$$
$$.05x + .10x + .30 = 2.10$$
$$.15x + .30 = 2.10$$
$$.15x = 1.80$$
$$x = 12$$

There are 12 nickels and 15 dimes.

45.

	Invested at 10%	Invested at 8%
Number of	2x	x
Interest on	.10(2x)	.08x

$$.10(2x) + .08x = 224$$
$$20x + .08x = 224$$
$$.28x = 224$$
$$x = 800$$

The man invested $800 at 8% and $1,600 at 10%.

Section 3.9

1. Step 1: Let x represent one of the numbers and y represent the other.

    Step 2: Two numbers have a sum of 25.

    $$x + y = 25$$

    One number is five more than the other.

    $$x = y + 5$$

    Step 3: Substitute y + 5 for x into the first equation.

    $$(y + 5) + y = 25$$
    $$2y + 5 = 25$$
    $$2y = 20$$
    $$y = 10$$

    When y = 10,   x = y + 5
    becomes   x = 10 + 5 = 15

    Step 4: The numbers are 10 and 15.

5. Step 1: Let x represent one of the positive numbers and y represent the other.

    Step 2: Two positive numbers have a difference of 5.

    $$x - y = 5$$

    The larger number is one more than twice the smaller.

    $$x = 2y + 1$$

    Step 3: Substitute 2y + 1 for x in the first equation.

    $$(2y + 1) - y = 5$$
    $$y + 1 = 5$$
    $$y = 4$$

    When y = 4,   x = 2y + 1
    becomes   x = 2(4) + 1 = 8 + 1 = 9

    Step 4: The largest number is 9 and the smaller number is 4.

Section 3.9 continued

9. Step 1: Let x = amount invested at 6% and y = amount invested at 8%.

   Step 2: His total investment is $20,000.

   $$x + y = \$20,000$$

   $$\underbrace{\text{Interest at 6\%}} + \underbrace{\text{Interest at 8\%}} = \underbrace{\text{Total Interest}}$$

   $$0.06x + 0.08y = 1380$$

   Step 3:
   $$x + y = 20,000 \xrightarrow{\text{Multiply by -6}} -6x - 6y = -120,000$$
   $$0.06x + 0.08y = 1,380 \xrightarrow{\text{Multiply by 100}} \underline{6x + 8y = 138,000}$$
   $$2y = 18,000$$
   $$y = 9,000$$

   Substitute y = 9,000 into the first equation.

   $$x + 9,000 = 20,000$$
   $$x = 11,000$$

   Step 4: Mr. Wilson invested $11,000 at 6% and $9,000 at 8%.

13. Step 1: Let x = the number of nickels and y = the number of quarters.

    Step 2: The total number of coins is 14.

    $$x + y = 14$$

    $$\underbrace{\text{Amount of money in nickels}} + \underbrace{\text{Amount of money in quarters}} = \underbrace{\text{Total amount of money}}$$

    $$.05x + .25y = 2.30$$

    Step 3:
    $$x + y = 14 \xrightarrow{\text{Multiply by -5}} -5x - 5y = -70$$
    $$.05x + .25y = 2.30 \xrightarrow{\text{Multiply by 100}} \underline{5x + 25y = 230}$$
    $$20y = 160$$
    $$y = 8$$

    Substituting y = 8 in the first equation

    $$x + 8 = 14$$
    $$x = 6$$

    Step 4: Ron has 6 nickels and 8 quarters.

17. Step 1: Let x = the number of liters of 50% alcohol solution and let y = the number of liters of 20% solution.

Step 2: Total number of liters is 18.

$$x + y = 18$$

$$\underbrace{50\% \text{ solution}}_{\text{of } x} + \underbrace{20\% \text{ solution}}_{\text{of } y} = \underbrace{30\% \text{ solution}}_{\text{of } 18 \text{ liters}}$$

$$.50x \qquad .20y \qquad = \qquad .30(18)$$

Step 3:
$$x + y = 18$$
$$.50x + .20y = .30(18)$$

Multiply by -20 →

Multiply by 100 →

$$-20x - 20y = -360$$
$$\underline{50x + 20y = 540}$$
$$30x = 180$$
$$x = 6$$

Substitute x = 6 into the first equation.

$$6 + y = 18$$
$$y = 12$$

Step 4: It takes 6 liters of 50% solution and 12 liters of 20% solution to produce 18 liters of 30% alcohol solution.

21. Let x = adult tickets and y = kids under 12 tickets.

$$x + y = 70 \qquad \text{The number of tickets}$$
$$5.50x + 4.00y = 310.00 \qquad \text{The value of the tickets}$$

multiply by -400 →

multiply by 100 →

$$x + y = 70$$
$$5.50x + 4.00y = 310.00$$

$$-400x - 400y = -28,000$$
$$\underline{550x + 400y = 31,000}$$
$$150x = 3,000$$
$$x = 20$$

Substituting x = 20 into our first equation, we get y = 50.

There were 20 adult tickets sold and 50 kids under 12 tickets sold.

25. Let x = $5.00 chips and y = $25.00 chips.

$$x + y = 45 \qquad \text{The number of chips}$$
$$5x + 25y = 465 \qquad \text{The value of the chips}$$

multiply by -5 →

no change →

$$x + y = 45$$
$$5x + 25y = 465$$

$$-5x - 5y = -225$$
$$\underline{5x + 25y = 465}$$
$$20y = 240$$
$$y = 12$$

Substituting y = 12 into our first equation, we get x = 33.

The gambler finished with 33 $5.00 chips and 12 $25.00 chips.

Section 3.9 continued

29. $2x - 6 > 5x + 6$

$2x - 12 > 5x$ Add -6 to both sides

$-12 > 3x$ Add -2x to both sides

$-4 > x$ Multiply each side by $\frac{1}{3}$

$x < -4$ Rewritten

33. $4(2 - x) \geq 12$

$8 - 4x \geq 12$ Distributive property

$-4x \geq 4$ Add -8 to both sides

$x \leq -1$ Multiply by $-\frac{1}{4}$ and reverse the direction of the inequality symbol.

91

Chapter 3 Review

1. $3x + y = 6$ $(4,-6)$ $3x + y = 6$ $(0,6)$
        $3(4) + y = 6$ $3(0) + y = 6$
        $12 + y = 6$ $y = 6$
              $y = -6$

        $3x + y = 6$ $(1,3)$ $3x + y = 6$ $(2,0)$
        $3x + 3 = 6$ $3x + 0 = 6$
            $3x = 3$ $3x = 6$
              $x = 1$ $x = 2$

5. $y = -3$ $(2,-3)$ $(-1,-3)$ $(-3,-3)$

9. $y = 3x + 7$ $(-\frac{8}{3}, -1)$ is the solution

     $-1 = 3(-\frac{8}{3}) + 7$

     $-1 = -8 + 7$

     $-1 = -1$ True

13.   See the graph on page A-16 in the textbook.

17. $x + y = -2$ $(-2,0)$ $x + y = -2$ $(0,-2)$
     $x + 0 = -2$ $0 + y = -2$
        $x = -2$ $y = -2$

     $x + y = -2$ $(1,-3)$
     $1 + y = -2$
        $y = -3$

   See the graph on page A-16 in the textbook.

21. $y = 2x - 1$ $(1,1)$ $y = 2x - 1$ $(0,-1)$
     $y = 2(1) - 1$ $y = 2(0) - 1$
     $y = 2 - 1$ $y = -1$
     $y = 1$

     $y = 2x - 1$ $(-1,-3)$
     $-3 = 2x - 1$
     $-2 = 2x$
     $-1 = x$

   See the graph on page A-16 in the textbook.

Chapter 3 Review continued

25. **y-intercept**

If $\qquad$ x = 0 $\qquad\qquad$ If $\qquad$ x = 3

the equation $\quad y = -\frac{1}{3}x$ $\qquad$ the equation $\quad y = -\frac{1}{3}x$

becomes $\qquad y = -\frac{1}{3}(0)$ $\qquad$ becomes $\qquad y = -\frac{1}{3}(3)$

$\qquad\qquad\qquad y = 0$ $\qquad\qquad\qquad\qquad y = -1$

$\qquad\qquad\qquad (0,0)$ $\qquad\qquad\qquad\qquad (3,-1)$

See the graph on page A-17 in the textbook.

29. See the graph on page A-17 in the textbook.

33. **x-intercept** $\qquad\qquad\qquad\qquad$ **y-intercept**

If $\qquad\qquad\qquad y = 0$ $\qquad\qquad$ If $\qquad\qquad\qquad x = 0$
the equation $\quad 3x - y = 6$ $\qquad$ the equation $\quad 3x - y = 6$
becomes $\qquad 3x - 0 = 6$ $\qquad$ becomes $\qquad 3(0) - y = 6$
$\qquad\qquad\qquad 3x = 6$ $\qquad\qquad\qquad\qquad -y = 6$
$\qquad\qquad\qquad x = 2$ $\qquad\qquad\qquad\qquad y = -6$
$\qquad\qquad\qquad (2,0)$ $\qquad\qquad\qquad\qquad (0,-6)$

37. **x-intercept** $\qquad\qquad\qquad\qquad$ **y-intercept**

If $\qquad\qquad\qquad y = 0$ $\qquad\qquad$ If $\qquad\qquad\qquad x = 0$
the equation $\quad y = 3x - 6$ $\qquad$ the equation $\quad y = 3x - 6$
becomes $\qquad 0 = 3x - 6$ $\qquad$ becomes $\qquad y = 3(0) - 6$
$\qquad\qquad\qquad 6 = 3x$ $\qquad\qquad\qquad\qquad y = -6$
$\qquad\qquad\qquad 2 = x$ $\qquad\qquad\qquad\qquad (0,-6)$
$\qquad\qquad\qquad (2,0)$

41. Let $(x_1,y_1) = (-2,3)$ and $(x_2,y_2) = (6,-5)$

$$m = \frac{y_2 - y_1}{x_2 - x_1} = \frac{-5 - 3}{6 - (-2)} = \frac{-8}{8} = -1$$

45. If $y = mx + b$, substituting $m = 3$, $b = 2$

$\qquad y = 3x + 2$

49. If $y = mx + b$, substituting $-\frac{1}{3}$, $b = \frac{3}{4}$

$\qquad y = -\frac{1}{3}x + \frac{3}{4}$

53. $2x + y = -5$
$\qquad y = -2x - 5$
$\qquad m = -2$, $b = -5$

57. $5x + 2y = 8$

$$2y = -5x + 8$$

$$y = \frac{-5x}{2} + \frac{8}{2}$$

$$y = -\frac{5}{2}x + 4$$

$$m = -\frac{5}{2}, \; b = 4$$

61. $y = mx + b$

$$3 = \frac{1}{2}(4) + b$$

$$3 = 2 + b$$

$$1 = b$$

If $m = \frac{1}{2}$ and $b = 1$, then the equation is $y = \frac{1}{2}x + 1$.

65. Let $(x_1, y_1) = (2,4)$ and $(x_2, y_2) = (1,6)$

$$m = \frac{y_2 - y_1}{x_2 - x_1} = \frac{6 - 4}{1 - 2} = \frac{2}{-1} = -2$$

using $\quad (x_1, y_1) = (2,4)$ and $m = -2$

in $\qquad y - y_1 = m(x - x_1)$ $\qquad\qquad$ Point-slope form

gives us $\quad y - 4 = -2(x - 2)$
$$y - 4 = -2x + 4$$
$$y = -2x + 8$$

69. Let $(x_1, y_1) = (3,-7)$ and $(x_2, y_2) = (-3,1)$

$$m = \frac{y_2 - y_1}{x_2 - x_1} = \frac{1 - (-7)}{-3 - 3} = \frac{8}{-6} = -\frac{4}{3}$$

using $\quad (x_1, y_1) = (3,-7)$ and $m = -\frac{4}{3}$

in $\qquad y - y_1 = m(x - x_1)$ $\qquad\qquad$ Point-slope form

gives us $\quad y - (-7) = -\frac{4}{3}(x - 3)$

$$y + 7 = -\frac{4}{3}x + 4$$

$$y = -\frac{4}{3}x - 3$$

73.

x-intercept		**y-intercept**	
When	$y = 0$	When	$x = 0$
the equation	$2x - 3y = 12$	the equation	$2x - 3y = 12$
becomes	$2x - 3(0) = 12$	becomes	$2(0) - 3y = 12$
	$2x = 12$		$-3y = 12$
	$x = 6$		$y = -4$
	$(6,0)$		$(0,-4)$

x-intercept		**y-intercept**	
When	$y = 0$	When	$x = 0$
the equation	$-2x + y = -8$	the equation	$-2x + y = -8$
becomes	$-2x + 0 = -8$	becomes	$-2(0) + y = -8$
	$-2x = -8$		$y = -8$
	$x = 4$		
	$(4,0)$		$(0,-8)$

See the graph on page A-18 in the textbook.

77. $x - y = 4$ Substituting $x = 1$ in the second equation,

$$\underline{x + y = -2} \qquad 1 + y = -2$$
$$2x = 2 \qquad\qquad y = -3$$
$$x = 1$$

The ordered pair is $(1,-3)$.

81.

multiply by 4

$-3x + 4y = 1 \xrightarrow{} -12x + 16y = 4$

$-4x + 5y = 1 \xrightarrow{} \underline{12x - 15y = -3}$

multiply by -3 $\qquad\qquad\qquad y = 1$

Substituting $y = 1$ in the first equation,

$$-3x + 4(1) = 1$$
$$-3x + 4 = 1$$
$$-3x = -3$$
$$x = 1$$

The ordered pair is $(1,1)$.

85. $x + y = 5$ and $y = -3x + 1$

$$x + (-3x + 1) = 5$$
$$-2x + 1 = 5$$
$$-2x = 4$$
$$x = -2$$

Substituting $x = -2$ in the first equation,

$$-2 + y = 5$$
$$y = 7$$

The ordered pair is $(-2,7)$.

Chapter 3 Review continued

89. $-3x + 12y = -8$ and $x = 4y + 2$

$$-3(4y + 2) + 12y = -8$$
$$-12y - 6 + 12y = -8$$
$$-6 = -8 \qquad \text{False statement}$$

There are no points in common, therefore the lines are parallel.

93. Let x = smaller number and y = larger number

$$x + y = 18$$
$$2x = y + 6 \quad \text{becomes} \quad 2x - y = 6$$

$$\begin{array}{r} x + y = 18 \\ 2x - y = 6 \\ \hline 3x = 24 \\ x = 8 \end{array}$$

Substituting $x = 8$ in the first equation,

$$8 + y = 18$$
$$y = 10$$

The ordered pair is (8,10).

97. Let x = dimes and y = nickels

$$x + y = 17 \qquad \text{The number of coins}$$
$$0.10x + 0.05y = 1.35 \qquad \text{The value of the coins}$$

$$\begin{array}{l} x + y = 17 \\ 0.10x + 0.05 = 1.35 \end{array} \xrightarrow[\text{multiply by 100}]{\text{multiply by -10}} \begin{array}{r} -10x - 10y = -170 \\ 10x + 5y = 135 \\ \hline -5y = -35 \\ y = 7 \end{array}$$

Substituting $x = 7$ in the first equation

$$x + 7 = 17 \quad \text{so} \quad x = 10$$

Barbara has 10 dimes and 7 nickels.

Chapter 3 Test

1. Given the equation $2x - 5y = 10$

    Given $(0, \ )$, substitute $x = 0$

    $$2(0) - 5y = 10$$
    $$-5y = 10$$
    $$y = -2$$

    The ordered pair is $(0,-2)$.

    Given $(\ ,0)$, substitute $y = 0$

    $$2x - 5(0) = 10$$
    $$2x = 10$$
    $$x = 5$$

    The ordered pair is $(5,0)$.

    Given $(10, \ )$, substitute $x = 10$

    $$2(10) - 5y = 10$$
    $$20 - 5y = 10$$
    $$-5y = -10$$
    $$y = 2$$

    The ordered pair is $(10,2)$.

    Given $(\ ,-3)$, substitute $y = -3$

    $$2x - 5(-3) = 10$$
    $$2x + 15 = 10$$
    $$2x = -5$$
    $$x = -\frac{5}{2}$$

    The ordered pair is $(-\frac{5}{2},-3)$.

2. $y = 4x - 3$ $(2,5)$ a solution
    $5 = 4(2) - 3$
    $5 = 8 - 3$
    $5 = 5$ True

    $y = 4x - 3$ $(0,-3)$ a solution
    $-3 = 4(0) - 3$
    $-3 = -3$ True

    $y = 4x - 3$ $(3,0)$ not a solution
    $0 = 4(3) - 3$
    $0 = 12 - 3$
    $0 = 9$ False

    $y = 4x - 3$ $(-2,11)$ not a solution
    $11 = 4(-2) - 3$
    $11 = -11$ False

3. $y = 3x - 2$

x-intercept

When	$y = 0$
the equation	$y = 3x - 2$
becomes	$0 = 3x - 2$
	$2 = 3x$
	$\frac{3}{2} = x$
	$(\frac{3}{2}, 0)$

y-intercept

When	$x = 0$
the equation	$y = 3x - 2$
becomes	$y = 3(0) - 2$
	$y = -2$
	$(0, -2)$

See the graph on page A-18 in the textbook.

4. $x = -2$ y is all real numbers

See the graph on page A-18 in the textbook.

5. $3x - 2y = 6$

$3x - 2y = 6$ into $y = mx + b$
$-2y = -3x + 6$

$y = \frac{3}{2}x - 3$

slope = $m = \frac{3}{2}$ y-intercept = $b = -3$

x-intercept when $y = 0$

$3x - 2(0) = 6$
$3x = 6$
$x = 2$ x-intercept is 2

See the graph on page A-18 in the textbook.

6. $2x - y = 5$ in $y = mx + b$

$-y = -2x + 5$
$y = 2x - 5$

slope = $m = 2$ y-intercept = $b = -5$

x-intercept when $y = 0$

$2x - y = 5$
$2x - 0 = 5$
$2x = 5$

$x = \frac{5}{2}$ is x-intercept

See the graph on page A-18 in the textbook.

7. Slope is 4   y-intercept is 8

$$y = mx + b$$
$$y = 4x + 8$$

8. Let $(x_1, y_1) = (-3, 1)$ and $(x_2, y_2) = (-2, 4)$

$$m = \frac{y_2 - y_1}{x_2 - x_1} = \frac{4 - 1}{-2 - (-3)} = \frac{3}{1} = 3$$

(You may substitute either set of points.)

$$y = mx + b \qquad x = -3, \ y = 1, \ m = 3$$
$$1 = +3(-3) + b$$
$$1 = -9 + b$$
$$10 = b$$

The equation of the line is $y = 3x + 10$

9. $x + 2y = 5$

x-intercept   **y-intercept**

When $y = 0$ When $x = 0$
the equation $x + 2y = 5$ the equation $x + 2y = 5$
becomes $x + 2(0) = 5$ becomes $0 + 2y = 5$
                           $x = 5$ $y = \frac{5}{2}$
                           $(5,0)$

                                                                              $(0, \frac{5}{2})$

$y = 2x$
x-intercept
When $y = 0$ When $x = 1$
the equation $y = 2x$ the equation $y = 2x$
becomes $0 = 2x$ becomes $y = 2(1)$
                           $0 = x$ $y = 2$
                           $(0,0)$ $(1,2)$

See the graph on page A-18 in the textbook.

10.
$$
\begin{array}{l}
x - y = 1 \xrightarrow{\text{multiply by } -2} -2x + 2y = -2 \\
2x - 3y = 6 \xrightarrow{\text{no change}} \underline{2x - 3y = 6} \\
\phantom{2x - 3y = 6 \xrightarrow{\text{no change}}} \ \ -y = 4 \\
\phantom{2x - 3y = 6 \xrightarrow{\text{no change}}} \ \ \ \ y = -4
\end{array}
$$

Substituting $y = -4$ in the first equation,

$$x - (-4) = 1$$
$$x + 4 = 1$$
$$x = -3$$

The solution to the system is $(-3, -4)$.

Chapter 3 Test continued

11.

$$2x + y = 7 \xrightarrow{\text{multiply by -7}} -14x - 7y = -49$$
$$3x + 7y = -6 \xrightarrow{\text{no change}} \underline{\quad 3x + 7y = -6}$$
$$-11x = -55$$
$$x = 5$$

$$2x + y = 7 \qquad \text{when } x = 5$$
$$2(5) + y = 7$$
$$10 + y = 7$$
$$y = -3$$

The solution to the system is (5,-3).

12.

$$7x + 8y = -2 \xrightarrow{\text{no change}} 7x + 8y = -2$$
$$3x - 2y = 10 \xrightarrow{\text{multiply by 4}} \underline{\quad 12x - 8y = 40}$$
$$19x = 38$$
$$x = 2$$

$$7x + 8y = -2 \qquad \text{when } x = 2$$
$$7(2) + 8y = -2$$
$$14 + 8y = -2$$
$$8y = -16$$
$$y = -2$$

The solution to the system is (2,-2).

13.

$$6x - 10y = 6 \xrightarrow{\text{multiply by -3}} 18x - 30y = 18$$
$$9x - 15y = 9 \xrightarrow{\text{multiply by -2}} \underline{\quad -18x + 30y = -18}$$
$$0 = 0$$

The lines coincide and there is an infinite number of solutions to the system.

14. $3x + 2y = 20$
   $y = 2x + 3$

Substituting 2x + 3 for y in the first equation,

$$3x + 2(2x + 3) = 20$$
$$3x + 4x + 6 = 20$$
$$7x + 6 = 20$$
$$7x = 14$$
$$x = 2$$

$$y = 2x + 3 \qquad \text{when } x = 2$$
$$y = 2(2) + 3$$
$$y = 4 + 3$$
$$y = 7$$

The solution to the system is (2,7).

15. $3x - 6y = -6$
$x = 3$

Substituting $x = 3$ for x in the first equation.

$3(3) - 6y = -6$
$9 - 6y = -6$
$-6y = -15$

$$y = \frac{-15}{-6}$$

$$y = \frac{5}{2}$$

The solution to the system is $(3, \frac{5}{2})$

16. $2x - 7y = 2$
$x = 4y$

Substituting $4y$ for x in the first equation,

$2(4y) - 7y = 2$
$8y - 7y = 2$
$y = 2$

$x = 2y$ when $y = 2$
$x = 4(2)$
$x = 8$

The solution to the system is $(8, 2)$.

17. Step 1: Let $x =$ one number and $y =$ the other number.

Step 2: "The sum of the number is 12."

$x + y = 12$

"Their difference is 2."

$x - y = 2$

Step 3: $x + y = 12$
$\underline{x + y = 2}$
$2x = 14$
$x = 7$

$x + y = 12$ when $x = 7$
$7 + y = 12$
$y = 5$

Step 4: The numbers are 7 and 5.

18. Step 1: Let x = one number and y = the other number.

Step 2: "The sum of two number is 15."

$$x + y = 15$$

"One number is six more than twice the other."

$$x = 2y + 6$$

Step 3: $x + y = 15$
$x = 2y + 6$

Substitute 2y + 6 for x in the first equation.

$$2y + 6 + y = 15$$
$$3y + 6 = 15$$
$$3y = 9$$
$$y = 3$$

$x = 2y + 6$ when y = 3
$x = 2(3) + 6$
$x = 6 + 6$
$x = 12$

Step 4: The numbers are 12 and 3.

19. Step 1: Let x = amount invested at 9% and y = amount invested at 11%.

Step 2: "Total invested is $10,000."

$$x + y = 10,000$$

Interest at 9%	+	Interest at 11%	=	Total Interest
.09x	+	.11y	=	$980

Step 3:

$$x + y = 10,000 \xrightarrow{\text{multiply by -9}} -9x - 9y = -90,000$$
$$.09x + .11y = 980 \xrightarrow{\text{multiply by 100}} \underline{9x + 11y = 98,000}$$
$$2y = 98,000$$
$$y = 4,000$$

$x + y = 10,000$ when y = 4,000
$x + 4,000 = 10,000$
$x = 6,000$

Step 4: Dr. Stork invested $6,000 at 9% and $4,000 at 11%.

Chapter 3 Test continued

20.   Step 1: Let x = nickels and y = quarters.

      Step 2: Total coins are 12.

$$x + y = 12$$

Amount of money in nickels	+	Amount of money in quarters	=	Total amount of money
.05x	+	.25y	=	1.60

Step 3:

$$
\begin{array}{l}
x + y = 12 \\
.05x + .25y = 1.60
\end{array}
$$

multiply by -5 $\longrightarrow$ $-5x - 5y = -160$

multiply by 100 $\longrightarrow$ $5x + 25y = 160$

$$
\begin{array}{r}
20y = 100 \\
y = 5
\end{array}
$$

$$
\begin{array}{ll}
x + y = 12 & \text{when } y = 5 \\
x + 5 = 12 & \\
x = 7 &
\end{array}
$$

Step 4: Diane has 7 nickels and 5 quarters.

Chapter 4

Section 4.1

1. $4^2 = 4 \cdot 4 = 16$ Base 4, exponent 2

5. $4^3 = 4 \cdot 4 \cdot 4 = 16 \cdot 4 = 64$ Base 4, exponent 3

9. $-2^3 = -2 \cdot 2 \cdot 2 = -4 \cdot 2 = -8$ Base 2, exponent 3

13. $\left(\frac{2}{3}\right)^2 = \frac{2}{3} \cdot \frac{2}{3} = \frac{4}{9}$ Base $\frac{2}{3}$, exponent 2

17. $x^4 \cdot x^5 = x^{4+5} = x^9$

21. $y^{10} \cdot y^{20} = y^{10+20} = y^{30}$

25. $x^4 \cdot x^6 \cdot x^8 \cdot x^{10} = x^{4+6+8+10} = x^{28}$

29. $\left(5^4\right)^3 = 5^{4 \cdot 3} = 5^{12}$

33. $\left(2^5\right)^{10} = 2^{5 \cdot 10} = 2^{50}$

37. $\left(b^x\right)^y = b^{xy}$

41. $(2y)^5 = 2^5 y^5$

$\qquad = 32y^5$

45. $(.5ab)^2 = .5^2 a^2 b^2$

$\qquad = .25 a^2 b^2$

49. $\left(2x^4\right)^3 = 2^3 x^{4 \cdot 3}$

$\qquad = 2^3 x^{12}$

$\qquad = 8x^{12}$

53. $\left(x^2\right)^3 \left(x^4\right)^2 = x^{2 \cdot 3} x^{4 \cdot 2}$

$\qquad = x^6 x^8$

$\qquad = x^{6+8}$

$\qquad = x^{14}$

57. $(2x)^3(2x)^4 = 2^3x^3 2^4x^4$

$\qquad = 2^3 \cdot 2^4 x^3 x^4$

$\qquad = 2^{3+4} x^{3+4}$

$\qquad = 2^7 x^7$

$\qquad = 128 x^7$

61. $(4x^2y^3)^2 = 4^2 x^{2 \cdot 2} y^{3 \cdot 2}$

$\qquad = 4^2 x^4 y^6$

$\qquad = 16 x^4 y^6$

65. $(3x^2)(2x^3)(5x^4) = 3 \cdot 2 \cdot 5 \cdot x^2 x^3 x^4$

$\qquad = 30 x^{2+3+4}$

$\qquad = 30 x^9$

69. $43{,}200 = 4.32 \times 10^4$

73. $238{,}000 = 2.38 \times 10^5$

77. $3.52 \times 10^2 = 3.52 \times 100$

$\qquad = 352$

81. $x^m x^m = x^{m+m} = x^{2m}$

85. $(x^{3m})^2 (x^{4m})^3 = x^{3m \cdot 2} x^{4m \cdot 3}$

$\qquad = x^{6m} x^{12m}$

$\qquad = x^{6m+12m}$

$\qquad = x^{18m}$

89. $7.4 \times 10^5 = 7.4 \times 100{,}000$

$\qquad = 740{,}000$

93. $(2^3)^2 = 2^{3 \cdot 2} = 2^6$

$\qquad 2^{3^2} = 2^9 \qquad$ Note: $3^2 = 9$

Section 4.1 continued

97. $y = 2x + 2$ $(-2,-2)$ $y = 2x + 2$ $(0,2)$
   $y = 2(-2) + 2$ $y = 2(0) + 2$
   $y = -4 + 2$ $y = 2$
   $y = -2$

   $y = 2x + 2$ $(2,6)$
   $y = 2(2) + 2$
   $y = 4 + 2$
   $y = 6$

See the graph on page A-19 in the textbook.

Section 4.2

1. $3^{-2} = \dfrac{1}{3^2}$ Definition of negative exponents

   $= \dfrac{1}{9}$

5. $8^{-2} = \dfrac{1}{8^2}$ Definition of negative exponents

   $= \dfrac{1}{64}$

9. $2x^{-3} = 2\left(\dfrac{1}{x^3}\right)$ Definition of negative exponents

   $= \dfrac{2}{x^3}$

13. $(5y)^{-2} = \dfrac{1}{(5y)^2}$ Definition of negative exponents

   $= \dfrac{1}{5^2 y^2}$ Property 3

   $= \dfrac{1}{25y^2}$

17. $\dfrac{5^3}{5^1} = 5^{3-1}$ Property 4

   $= 5^2$
   $= 25$

21. $\dfrac{x^{10}}{x^4} = x^{10-4}$ Property 4

   $= x^6$

25. $\dfrac{6^{11}}{6} = \dfrac{6^{11}}{6^1}$ Definition

     $= 6^{11-1}$ Property 4

     $= 6^{10}$ Subtraction

29. $\dfrac{2^{-5}}{2^3} = 2^{-5-3}$ Property 4

     $= 2^{-8}$

     $= \dfrac{1}{2^8}$ Definition of negative exponents

33. $(2x^{-3})^4 = 2^4 x^{-3 \cdot 4}$ Property 3

     $= 2^4 x^{-12}$ Property 2

     $= \dfrac{16}{x^{12}}$

37. $(2a^2 b)^1 = 2a^2 b$ Property 3

41. $x^{-3} x^{-5} = x^{-3(-5)}$ Property 1

     $= x^{-8}$

     $= \dfrac{1}{x^8}$ Definition of negative exponents

45. $\dfrac{(a^4)^3}{(a^3)^2} = \dfrac{a^{12}}{a^6}$ Property 2

     $= a^{12-6}$ Property 4

     $= a^6$

49. $\left(\dfrac{y^7}{y^2}\right)^8 = (y^{7-2})$ Property 4

     $= (y^5)^8$

     $= y^{40}$ Property 2

53. $\left(\dfrac{x^{-2}}{x^{-5}}\right)^3 = \dfrac{x^{-6}}{x^{-15}}$ Property 5

     $= x^{-6-(-15)}$ Property 4

     $= x^9$ Subtraction

Section 4.2 continued

57. $\dfrac{(a^{-2})^3(a^4)^2}{(a^{-3})^{-2}} = \dfrac{a^{-6}a^8}{a^6}$ Property 2

$\qquad\qquad\qquad = \dfrac{a^2}{a^6}$ Property 1

$\qquad\qquad\qquad = a^{2-6}$ Property 4

$\qquad\qquad\qquad = a^{-4}$ Subtraction

$\qquad\qquad\qquad = \dfrac{1}{a^4}$ Definition

61. $0.000357 = 3.57 \times 10^{-4}$

65. $0.0048 = 4.8 \times 10^{-3}$

69. $4.23 \times 10^{-3} = 0.00423$

73. $7.89 \times 10^1 = 78.9$

77. $\dfrac{x^{4m}}{x^m} = x^{4m-m}$ Property 4

$\qquad\quad = x^{3m}$ Subtraction

81. $\dfrac{x^{-4m}}{x^{-9m}} = x^{-4m-(-9m)}$ Property 4

$\qquad\quad = x^{5m}$

85. $0.006 = 6.0 \times 10^{-3}$

89. $4x + 3x = (4 + 3)x$
$\qquad\qquad = 7x$

93. $4y + 5y + y = (4 + 5 + 1)y$
$\qquad\qquad\quad = 10y$

97.

x-intercept			y-intercept		
When		$y = 0$	When		$x = 0$
the equation		$2x + y = 4$	the equation		$2x + y = 4$
becomes		$2x + 0 = 4$	becomes		$2(0) + y = 4$
		$2x = 4$			$y = 4$
		$x = 2$			$(0,4)$
		$(2,0)$			

See the graph on page A-19 in the textbook.

Section 4.3

1. $(3x^4)(4x^3) = (3 \cdot 4)(x^4 \cdot x^3)$ Commutative and associative properties

        $= 12x^7$ Multiply coefficients, add exponents

5. $(8x)(4x) = (8 \cdot 4)(x \cdot x)$ Commutative and associative properties

        $= 32x^2$ Multiply coefficients, add exponents

9. $(6ab^2)(-4a^2b) = 6(-4)(aa^2)(b^2b)$ Commutative and associative properties

               $= -24a^3b^3$ Multiply coefficients, add exponents

13. $\dfrac{15x^3}{5x^2} = \dfrac{15}{5} \cdot \dfrac{x^3}{x^2}$ Write as separate fractions

        $= 3x$ Divide coefficients, subtract exponents

17. $\dfrac{32a^3}{64a^4} = \dfrac{32}{64} \cdot \dfrac{a^3}{a^4}$ Write as separate fractions

        $= \dfrac{1}{2} \cdot \dfrac{1}{a}$ Divide coefficients, subtract exponents

        $= \dfrac{1}{2a}$

21. $\dfrac{3x^3y^2z}{27xy^2z^3} = \dfrac{3}{27} \cdot \dfrac{x^3}{x} \cdot \dfrac{y^2}{y^2} \cdot \dfrac{z}{z^3}$ Write as separate fractions

        $= \dfrac{1}{9} \cdot \dfrac{x^2}{1} \cdot \dfrac{1}{z^2}$ Divide coefficients, subtract exponents

        $= \dfrac{x^2}{9z^2}$

25. $(3 \times 10^3)(2 \times 10^3) = (3 \times 2)(10^3 \times 10^5)$

                 $= 6 \times 10^8$

29. $(5.5 \times 10^{-3})(2.2 \times 10^{-4}) = (5.5 \times 2.2)(10^{-3} \times 10^{-4})$

                      $= 12.1 \times 10^{-7}$

                      $= 1.21 \times 10^1 \times 10^{-7}$

                      $= 1.21 \times 10^{-6}$

Section 4.3 continued

33. $\dfrac{6 \times 10^8}{2 \times 10^{-2}} = \dfrac{6}{2} \times \dfrac{10^8}{10^{-2}}$

    $= 3 \times 10^{8-(-2)}$

    $= 3 \times 10^{10}$

37. $3x^2 + 5x^2 = (3 + 5)x^2$

    $= 8x^2$

41. $2a + a - 3a = (2 + 1 - 3)a$

    $= 0a$

    $= 0$

45. $20ab^2 - 19ab^2 + 30ab^2 = (20 - 19 + 30)ab^2$

    $= 31ab^2$

49. $\dfrac{(3x^2)(8x^5)}{6x^4} = \left(\dfrac{3 \cdot 8}{6}\right)\dfrac{x^2 \cdot x^5}{x^4}$

    $= 4\dfrac{x^7}{x^4}$

    $= 4x^{7-4}$

    $= 4x^3$

53. $\dfrac{(4x^3y^2)(9x^4y^{10})}{(3x^5y)(2x^6y)} = \dfrac{4 \cdot 9}{3 \cdot 2} \cdot \dfrac{x^3yx^4}{x^5x^6} \cdot \dfrac{y^2y^{10}}{yy}$

    $= \dfrac{36}{6} \cdot \dfrac{x^7}{x^{11}} \cdot \dfrac{y^{12}}{y^2}$

    $= \dfrac{6y^{10}}{x^4}$

57. $\dfrac{(5 \times 10^3)(4 \times 10^{-5})}{2 \times 10} = \dfrac{5 \cdot 4}{2} \cdot \dfrac{10^3 \cdot 10^{-5}}{10^{-2}}$

    $= 10 \cdot \dfrac{10^{-2}}{10^{-2}}$

    $= 10 \cdot 10^0$

    $= 10$

    $= 1 \times 10^1$ Scientific Notation

61. $\dfrac{18x^4}{3x} + \dfrac{21x^7}{7x^4} = 6x^3 + 3x^3$

    $= 9x^3$

110

Section 4.3 continued

65. $\dfrac{6x^7y^4}{3x^2y^2} + \dfrac{8x^5y^8}{2y^6} = 2x^5y^2 + 4x^5y^2$

$$= 6x^3y^2$$

69. $(7^3)^X = 7^{12}$

$(7^3)^4 = 7^{12}$ Answer is x = 4.

73.

x	$y = 2x^2$	y
-3	$y = 2(-3)^2 = 2(9) = 18$	18
-2	$y = 2(-2)^2 = 2(4) = 8$	8
-1	$y = 2(-1)^2 = 2(1) = 2$	2
0	$y = 2(0)^2 = 2(0) = 0$	0
1	$y = 2(1)^2 = 2(1) = 2$	2
2	$y = 2(2)^2 = 2(4) = 8$	8
3	$y = 2(3)^2 = 2(9) = 18$	18

See the graph on page A-20 in the textbook.

77.

x	$y = \frac{1}{4}x^2$	y
-4	$y = \frac{1}{4}(-4)^2 = \frac{1}{4}(16) = 4$	4
-2	$y = \frac{1}{4}(-2)^2 = \frac{1}{4}(4) = 1$	1
-1	$y = \frac{1}{4}(-1)^2 = \frac{1}{4}(1) = \frac{1}{4}$	$\frac{1}{4}$
0	$y = \frac{1}{4}(0)^2 = \frac{1}{4}(0) = 0$	0
1	$y = \frac{1}{4}(1)^2 = \frac{1}{4}(1) = \frac{1}{4}$	$\frac{1}{4}$
2	$y = \frac{1}{4}(2) = \frac{1}{4}(4) = 1$	1
4	$y = \frac{1}{4}(4) = \frac{1}{4}(16) = 4$	4

See the graph on page A-20 in the textbook.

Section 4.3 continued

81. Substituting a = 5 and b = 4 in the expressions $(a + b)^2$ and $a^2 + 2ab + b^2$, we have

$$(a + b)^2 = a^2 + 2ab + b^2$$
$$(3 + 4)^2 = 3^2 + 2(3)(4) + 4^2$$
$$7^2 = 9 + 24 + 16$$
$$49 = 49$$

The expressions $(a + b)^2$ and $a^2 + 2ab + b^2$ are equal.

85. When x = -2, -2x + 5 becomes,
$$-2(-2) + 5 = 4 + 5 = 9$$

Section 4.4

1. $2x^3 - 3x^2 + 1$ Degree 3
     3 terms is a trinomial

5. $2x - 1$ Degree 1
     2 terms is a binomial

9. $7a^2$ Degree 2
     1 term is a monomial

13. $(2x^2 + 3x + 4) + (3x^2 + 2x + 5) = (2x^2 + 3x^2) + (3x + 2x) + (4 + 5)$
$$= 5x^2 + 5x + 9$$

17. $x^2 + 4x + 2x + 8 = x^2 + 6x + 8$

21. $x^2 - 3x + 3x - 9 = x^2 - 9$

25. $(6x^3 - 4x^2 + 2x) + (9x^2 - 6x + 3) = 6x^3 + (-4x^2 + 9x^2) + (2x - 6x) + 3$
$$= 6x^3 + 5x^2 - 4x + 3$$

29. $(a^2 - a - 1) - (-a^2 + a + 1) = a^2 - a - 1 + a^2 - a - 1$
$$= (a^2 + a^2) + (-a - a) + (-1 - 1)$$
$$= 2a^2 - 2a - 2$$

33. $(4y^2 - 3y + 2) + (5y^2 + 12y - 4) - (13y^2 - 6y + 20)$

$\quad = 4y^2 - 3y + 2 + 5y^2 + 12y - 4 - 13y^2 + 6y - 20$

$\quad = (4y^2 + 5y^2 - 13y^2) + (-3y + 12y + 6y) + (2 - 4 - 20)$

$\quad = -4y^2 + 15y - 22$

37. $(11y^2 + 11y + 11) - (3y^2 + 7y - 15)$

$\quad = 11y^2 + 11y + 11 - 3y^2 - 7y + 15$

$\quad = (11y^2 - 3y^2) + (11y - 7y) + (11 + 15)$

$\quad = 8y^2 + 4y + 26$

41. $[(3x - 2) + (11x + 5)] - (2x + 1)$

$\quad = [(3x + 11x) + (-2 + 5)] - (2x + 1)$

$\quad = 14x + 3 - 2x - 1$

$\quad = (14x - 2x) + (3 - 1)$

$\quad = 12x + 2$

45. When $\quad y = 10, \ (y - 5)^2$

becomes $(10 - 5)^2 = 5^2 = 25$

49.

x	$y = x^2 + 2$	y	
-3	$y = (-3)^2 + 2 = 9 + 2 = 11$	11	
-2	$y = (-2)^2 + 2 = 4 + 2 = 6$	6	
-1	$y = (-1)^2 + 2 = 1 + 2 = 3$	3	
0	$y = (0)^2 + 2 = 0 + 2 = 2$	2	vertex
1	$y = 1^2 + 2 \quad = 1 + 2 = 3$	3	
2	$y = 2^2 + 2 \quad = 4 + 2 = 6$	6	
3	$y = 3^2 + 2 \quad = 9 + 2 = 11$	11	

See the graph on page A-20 in the textbook.

53.

x	$y = x^2 + 3$	y	
-3	$y = (-3)^2 + 3 = 9 + 3 = 12$	12	
-2	$y = (-2)^2 + 3 = 4 + 3 = 7$	7	
-1	$y = (-1)^2 + 3 = 1 + 3 = 4$	4	
0	$y = 0^2 + 3 \quad = 0 + 3 = 3$	3	vertex
1	$y = 1^2 + 3 \quad = 1 + 3 = 4$	4	
2	$y = 2^2 + 3 \quad = 4 + 3 = 7$	7	
3	$y = 3^2 + 3 \quad = 9 + 3 = 12$	12	

See the graph on page A-20 in the textbook.

57.

x	$y = (x + 2)^2$	y	
-5	$y = (-5 + 2)^2 = (-3)^2 = 9$	9	
-4	$y = (-4 + 2)^2 = (-2)^2 = 4$	4	
-3	$y = (-3 + 2)^2 = (-1)^2 = 1$	1	
-2	$y = (-2 + 2)^2 = 0^2 \quad = 0$	0	vertex
-1	$y = (-1 + 2)^2 = 1^2 \quad = 1$	1	
0	$y = (0 + 2)^2 \quad = 2^2 \quad = 4$	4	
1	$y = (1 + 2)^2 \quad = 3^2 \quad = 9$	9	

See the graph on page A-20 in the textbook.

61. $2x(5x) = (2 \cdot 5)(x \cdot x)$
$$= 10x^2$$

65. $2x(3x^2) = (2 \cdot 3)(x \cdot x^2)$
$$= 6x^3$$

Section 4.5

1. $2x(3x + 1) = 2x(3x) + 2x(1)$
$$= 6x^2 + 2x$$

5. $2ab(a^2 - ab + 1) = 2ab(a^2) + 2ab(-ab) + 2ab(1)$
$$= 2a^3b - 2a^2b^2 + 2ab$$

9. $4x^2y(2x^3y + 3x^2y^2 + 8y^3) = 4x^2y(2x^3y) + 4x^2y(3x^2y^2) + 4x^2y(8y^3)$
$$= 8x^5y^2 + 12x^4y^3 + 32x^2y^4$$

13. $(x + 6)(x + 1) = x(x) + x(1) + 6(x) + 6(1)$
$$\qquad\qquad\quad F \qquad O \qquad I \qquad L$$
$$= x^2 + x + 6x + 6$$
$$= x^2 + 7x + 6$$

17. $(a + 5)(a - 3) = a(a) + a(-3) + 5(a) + 5(-3)$
$$\qquad\qquad\quad F \qquad O \qquad I \qquad L$$
$$= a^2 - 3a + 5a - 15$$
$$= a^2 + 2a - 15$$

21. $(x + 6)(x - 6) = x(x) + x(-6) + 6(x) + 6(-6)$

$$\qquad\qquad\qquad\quad\ \ \text{F}\qquad\ \ \text{O}\qquad\ \ \text{I}\qquad\ \ \text{L}$$

$$\qquad\qquad = \ \ x^2 \ \ - \ \ 6x \ \ + \ \ 6x \ \ - \ \ 36$$

$$\qquad\qquad = x^2 - 36$$

25. $(2x - 3)(x - 4) = 2x(x) + 2x(-4) + (-3)x + (-3)(-4)$

$$\qquad\qquad\qquad\qquad\ \ \text{F}\qquad\ \ \text{O}\qquad\quad\ \text{I}\qquad\quad\ \text{L}$$

$$\qquad\qquad = \ \ 2x^2 \ \ - \ \ 8x \ \ - \ \ 3x \ \ + \ \ 12$$

$$\qquad\qquad = 2x^2 - 11x + 12$$

29. $(2x - 5)(3x - 2) = 2x(3x) + 2x(-2) + (-5)3x + (-5)(-2)$

$$\qquad\qquad\qquad\qquad\ \ \text{F}\qquad\quad\ \text{O}\qquad\quad\ \text{I}\qquad\quad\ \text{L}$$

$$\qquad\qquad = \ \ 6x^2 \ \ - \ \ 4x \ \ - \ \ 15x \ \ + \ \ 10$$

$$\qquad\qquad = 6x^2 - 19x + 10$$

33. $(5x - 4)(5x + 4) = 5x(5x) + 5x(4) + (-4)5x + (-4)4$

$$\qquad\qquad\qquad\qquad\ \ \text{F}\qquad\quad\ \text{O}\qquad\quad\ \text{I}\qquad\quad\ \text{L}$$

$$\qquad\qquad = \ \ 25x^2 \ \ + \ \ 20x \ \ - \ \ 20x \ \ - \ \ 16$$

$$\qquad\qquad = 25x^2 - 16$$

37. $(1 - 2a)(3 - 4a) = 1(3) + 1(-4a) + (-2a)3 + (-2a)(-4a)$

$$\qquad\qquad\qquad\qquad\ \ \text{F}\qquad\ \ \text{O}\qquad\quad\ \text{I}\qquad\quad\ \ \text{L}$$

$$\qquad\qquad = \ \ 3 \ \ - \ \ 4a \ \ - \ \ 6a \ \ + \ \ 8a^2$$

$$\qquad\qquad = 3 - 10a + 8a^2$$

41. $(x + 1)(x^2 + 3x - 4)$

$$
\begin{array}{r}
x^2 + 3x - 4 \\
x + 1 \\
\hline
x^2 + 3x - 4 \\
x^3 + 3x^2 - 4x \qquad\quad \\
\hline
x^3 + 4x^2 - \ x - 4
\end{array}
$$

45. $(x + 2)(x^2 - 2x + 4)$

$$
\begin{array}{r}
x^2 - 2x + 4 \\
x + 2 \\
\hline
2x^2 - 4x + 8 \\
x^3 - 2x^2 + 4x \qquad\quad \\
\hline
x^3 \qquad\qquad\quad + 8
\end{array}
$$

Section 4.5 continued

49. $(5x^2 + 2x + 1)(x^2 - 3x + 5)$

$$
\begin{array}{r}
5x^2 + 2x + 1 \\
x^2 - 3x + 5 \\
\hline
25x^2 + 10x + 5 \\
-15x^3 - 6x^2 - 3x \\
5x^4 + 2x^3 + x^2 \\
\hline
5x^4 - 13x^3 + 20x^2 + 7x + 5
\end{array}
$$

61. If we let x = the width, then the length is $x + 1$. The area is

$A = L \times W$

$A = (x + 1)x$

$\quad = x^2 + x$

65. If we let x = number of calculators a company sells per day and p = the price of each calculator, the relationship between the calculators sold and the price of each calculator is given by the equation $x = 1700 - 100p$. If we let R = the daily revenues, we have

$\quad R = xp$

Substituting $x = 1700 - 100p$ in $R = xp$, we have

$\quad R = (1700 - 100p)p = 1700p - 100p^2$

69. $y = mx + b \qquad$ substituting $(x,y) = (-2,-6)$ and $m = 3$

$-6 = 3(-2) + b$

$-6 = -6 + b$

$\ \ 0 = b$

If $m = 3$ and $b = 0$, the equation is

$y = 3x + 0$

$y = 3x$

Section 4.6

1. $(x - 2)^2 = x^2 + 2(x)(-2) + 4$

$\qquad\qquad = x^2 - 4x + 4$

5. $(x - 5)^2 = x^2 + 2(x)(-5) + 25$

$\qquad\qquad = x^2 - 10x + 25$

9. $(x + 10)^2 = x^2 + 2(x)(10) + 100$

$\qquad\qquad\ = x^2 + 20x + 100$

13. $(2x - 1)^2 = 4x^2 + 2(2x)(-1) + 1$

$\qquad\qquad\quad = 4x^2 - 4x + 1$

17. $(3x - 2)^2 = 9x^2 + 2(3x)(-2) + 4$

$\qquad\qquad\quad = 9x^2 - 12x + 4$

21. $(4x - 5y)^2 = 16x^2 + 2(4x)(-5y) + 25y^2$

$\qquad\qquad\qquad = 16x^2 - 40xy + 25y^2$

25. $(6x - 10y)^2 = 36x^2 + 2(6x)(-10y) + 100y^2$

$\qquad\qquad\qquad = 36x^2 - 120xy + 100y^2$

29. $(a^2 + 1)^2 = a^4 + 2(a^2)(1) + 1^2$

$\qquad\qquad\quad = a^4 + 2a^2 + 1$

33. $(x - 3)(x + 3) = x^2 - 9$

37. $(y - 1)(y + 1) = y^2 - 1$

41. $(2x + 5)(2x - 5) = 4x^2 - 25$

45. $(2a + 7)(2a - 7) = 4a^2 - 49$

49. $(x^2 + 3)(x^2 - 3) = x^4 - 9$

53. $(5y^4 - 8)(5y^4 + 8) = 25y^8 - 64$

57. $(2x + 3)^2 - (4x - 1)^2$

$\qquad = (2x)^2 + 2(2x)(3) + 3^2 - [(4x)^2 + 2(4x)(-1) + (-1)^2]$

$\qquad = 4x^2 + 12x + 9 - (16x^2 - 8x + 1)$

$\qquad = 4x^2 + 12x + 9 - 16x^2 + 8x - 1$

$\qquad = -12x^2 + 20x + 8$

61. $(2x + 3)^3$

$\qquad = (2x + 3)^2(2x + 3)$

$\qquad = [4x^2 + 2(2x)(3) + 9](2x + 3)$

$\qquad = (4x^2 + 12x + 9)(2x + 3)$

$$
\begin{array}{r}
4x^2 + 12x + 9 \\
2x + 3 \\
\hline
12x^2 + 36x + 27 \\
8x^3 + 24x^2 + 18x \\
\hline
8x^3 + 36x^2 + 54x + 27
\end{array}
$$

$\qquad\qquad (2x + 3)^2 = 8x^3 + 36x^2 + 54x + 27$

65. $(x + 3)^2$ ___ substituting $x = 2$

$(2 + 3)^2 = 5^2 = 25$

$x^2 + 6x + 9$ ___ substituting $x = 2$

$$x^2 + 6x + 9$$
$$= 2^2 + 6(2) + 9$$
$$= 4 + 12 + 9$$
$$= 16 + 9$$
$$= 25$$

69. Let x ___ = first consecutive integer
Let $x + 1$ = second consecutive integer

$$x^2 + (x + 1)^2 = x^2 + x^2 + 2(x)(1) + 1^2$$
$$= x^2 + x^2 + 2x + 2$$
$$= 2x^2 + 2x + 2$$

73. $\dfrac{10x^3}{5x} = 2x^{3-1}$

$= 2x^2$

77. $\dfrac{35a^6b^8}{70a^2b^{10}} = \dfrac{a^{6-2}b^{8-10}}{2}$

$= \dfrac{a^4b^{-2}}{2}$

$= \dfrac{a^4}{2b^2}$

81. $y = 2x + 3$

x-intercept   **y-intercept**

When	$y = 0$	When	$x = 0$
the equation	$y = 2x + 3$	the equation	$y = 2x + 3$
becomes	$0 = 2x + 3$	becomes	$y = 2(0) + 3$

$\left(-\dfrac{3}{2}, 0\right)$ $-3 = 2x$ $(0, 3)$ $y = 3$

$-\dfrac{3}{2} = x$

$y = -2x - 1$

x-intercept   **y-intercept**

When	$y = 0$	When	$x = 0$
the equation	$y = -2x - 1$	the equation	$y = -2x - 1$
becomes	$0 = -2x - 1$	becomes	$y = -2(0) - 1$

$\left(-\dfrac{1}{2}, 0\right)$ $1 = -2x$ $(0, -1)$ $y = -1$

$-\dfrac{1}{2} = x$

The graph is on page A-22 in the textbook.

Section 4.7

1. $\dfrac{5x^2 - 10x}{5x} = \dfrac{5x^2}{5x} - \dfrac{10x}{5x}$

   $\qquad = x - 2$

5. $\dfrac{25x^2y - 10xy}{5x} = \dfrac{25x^2y}{5x} - \dfrac{10xy}{5x}$

   $\qquad = 5xy - 2y$

9. $\dfrac{50x^5 - 25x^3 + 5x}{5x} = \dfrac{50x^5}{5x} - \dfrac{25x^3}{5x} + \dfrac{5x}{5x}$

   $\qquad = 10x^4 - 5x^2 + 1$

13. $\dfrac{16a^5 + 24a^4}{-2a} = \dfrac{16a^5}{-2a} + \dfrac{24a^4}{-2a}$

    $\qquad = -8a^4 - 12a^3$

17. $\dfrac{12a^3b - 6a^2b^2 + 14ab^3}{-2a} = \dfrac{12a^3b}{-2a} - \dfrac{6a^2b^2}{-2a} + \dfrac{14ab^3}{-2a}$

    $\qquad = -6a^2b + 3ab^2 - 7b^3$

21. $\dfrac{6x + 8y}{2} = \dfrac{6x}{2} + \dfrac{8y}{2}$

    $\qquad = 3x + 4y$

25. $\dfrac{10xy - 8x}{2x} = \dfrac{10xy}{2x} - \dfrac{8x}{2x}$

    $\qquad = 5y - 4$

29. $\dfrac{x^2y - x^3y^2}{-x^2y} = \dfrac{x^2y}{-x^2y} - \dfrac{x^3y^2}{-x^2y}$

    $\qquad = -1 + xy$

33. $\dfrac{x^3 - 3x^2y + xy^2}{x} = \dfrac{x^3}{x} - \dfrac{3x^2y}{x} + \dfrac{xy^2}{x}$

    $\qquad = x^2 - 3xy + y^2$

37. $\dfrac{26x^2y^2 - 13xy}{-13xy} = \dfrac{26x^2y^2}{-13xy} - \dfrac{13xy}{-13xy}$

    $\qquad = -2xy + 1$

41. $\dfrac{5a^2x - 10ax^2 + 15a^2x^2}{20a^2x^2} = \dfrac{5a^2x}{20a^2x^2} - \dfrac{10ax^2}{20a^2x^2} + \dfrac{15a^2x^2}{20a^2x^2}$

$$= \frac{1}{4x} - \frac{1}{2a} + \frac{3}{4}$$

45. $\dfrac{9a^{5m} - 27a^{3m}}{3a^{2m}} = \dfrac{9a^{5m}}{3a^{2m}} - \dfrac{27a^{3m}}{3a^{2m}}$ Divide each term by $3a^{2m}$

$$= 3a^{5m-2m} - 9a^{3m-2m}$$ Simplify

$$= 3a^{3m} - 9a^{m}$$

49. $\dfrac{2x^3(3x + 2) - 3x^2(2x - 4)}{2x^2}$

$= \dfrac{6x^4 + 4x^3 - 6x^3 + 12x^2}{2x^2}$ Distributive property

$= \dfrac{6x^4}{2x^2} + \dfrac{4x^3}{2x^2} - \dfrac{6x^3}{2x^2} + \dfrac{12x^2}{2x^2}$ Divide each term by $2x^2$

$= 3x^2 + 2x - 3x + 6$ Simplify
$= 3x^2 - x + 6$ Simplify

53. $\dfrac{(x + 5)^2 + (x + 5)(x - 5)}{2x}$

$= \dfrac{x^2 + 10x + 25 + x^2 - 25}{2x}$ Multiply

$= \dfrac{2x^2 + 10x}{2x}$ Simplify

$= \dfrac{2x^2}{2x} + \dfrac{10x}{2x}$ Divide each term by $2x$

$= x + 5$

57. When $x = 10$

the expression $\dfrac{3x + 8}{2}$

becomes $\dfrac{3(10) + 8}{2} = \dfrac{30 + 8}{2}$

$= \dfrac{38}{2}$

$= 19$

When $x = 10$

the expression $3x + 4$

becomes $3(10) + 4 = 30 + 4$

$= 34$

Section 4.7 continued

61.

$$2x - 3y = -5 \xrightarrow{\text{no change}} 2x - 3y = -5$$

$$x + y = 5 \xrightarrow{\text{multiply by } -2} -2x - 2y = -10$$

$$\begin{array}{r} -5y = -15 \\ y = 3 \end{array}$$

Substituting $x = 3$ into $x + y = 5$ gives us

$$3 + y = 5$$
$$y = 2$$

The solution to the system is $(2,3)$.

65. Substituting $y = -2x + 4$ for y from the second equation into the first equation gives us

$$4x + 2(-2x + 4) = 8$$
$$4x - 4x + 8 = 8$$
$$8 = 8$$

A true statement, therefore the lines coincide.

Section 4.8

1.

$$\begin{array}{r}
x - 2 \\
x - 3 \overline{)\ x^2 - 5x + 6} \\
\underline{-\ \ \ \ +} \\
x^2 - 3x \\
\hline
-2x + 6 \\
\underline{+\ \ \ \ -} \\
-2x + 6 \\
\hline
0
\end{array}$$

Change signs

Change signs

Our answer: $x - 2$.

5.

$$\begin{array}{r}
x - 3 \\
x - 3 \overline{)\ x^2 - 6x + 9} \\
\underline{-\ \ \ \ +} \\
x^2 - 3x \\
\hline
-3x + 9 \\
\underline{+\ \ \ \ -} \\
-3x + 6 \\
\hline
0
\end{array}$$

Change signs

Change signs

Our answer: $x - 3$.

9.

$$\begin{array}{r}
a - 5 \\
2a + 1 \overline{)\ 2a^2 - 9a - 5} \\
\underline{-\ \ \ \ -} \\
2a^2 + a \\
\hline
-10a - 5 \\
\underline{+\ \ \ \ +} \\
-10a - 5 \\
\hline
0
\end{array}$$

Change signs

Change signs

Our answer: $a - 5$.

13.

$$
\begin{array}{r}
a - 2 \\
a + 5 \overline{\smash{)}\ a^2 + 3a + 2} \\
\end{array}
$$

Change signs

$$
\begin{array}{r}
- \quad - \\
a^2 + 5a \\
\hline
- 2a + 2 \\
\end{array}
$$

Change signs

$$
\begin{array}{r}
+ \quad + \\
- 2a - 10 \\
\hline
12 \\
\end{array}
$$

Our answer: $a - 2 + \dfrac{12}{a + 5}$

17.

$$
\begin{array}{r}
x + 4 \\
x + 1 \overline{\smash{)}\ x^2 + 5x - 6} \\
\end{array}
$$

Change signs

$$
\begin{array}{r}
- \quad - \\
x^2 + x \\
\hline
4x - 6 \\
\end{array}
$$

Change signs

$$
\begin{array}{r}
- \quad - \\
4x + 4 \\
\hline
- 10 \\
\end{array}
$$

Our answer: $x + 4 + \dfrac{-10}{x + 1}$

21.

$$
\begin{array}{r}
x - 3 \\
2x + 4 \overline{\smash{)}\ 2x^2 - 2x + 5} \\
\end{array}
$$

Change signs

$$
\begin{array}{r}
- \quad - \\
2x^2 + 4x \\
\hline
- 6x + 5 \\
\end{array}
$$

Change signs

$$
\begin{array}{r}
+ \quad + \\
- 6x - 12 \\
\hline
17 \\
\end{array}
$$

Our answer: $x - 3 + \dfrac{17}{2x + 4}$

25.

$$
\begin{array}{r}
2a^2 - a - 3 \\
3a - 5 \overline{\smash{)}\ 6a^3 - 13a^2 - 4a + 15} \\
\end{array}
$$

Change signs

$$
\begin{array}{r}
+ \quad + \\
6a^3 - 10a^2 \\
\hline
- 3a^2 - 4a \\
\end{array}
$$

Change signs

$$
\begin{array}{r}
+ \quad - \\
3a^2 + 5a \\
\hline
- 9a + 15 \\
\end{array}
$$

Change signs

$$
\begin{array}{r}
+ \quad - \\
9a + 15 \\
\hline
0 \\
\end{array}
$$

Our answer: $2a^2 - a - 3$

29.

$$x - 1 \overline{) \begin{array}{c} x^2 + x + 1 \\ x^3 + 0x^2 + 0x - 1 \end{array}}$$

Change signs

$$\begin{array}{c} - \quad + \\ \underline{x^3 - x^2} \\ x^2 + 0x \end{array}$$

Change signs

$$\begin{array}{c} - \quad + \\ \underline{x^2 - x} \\ x - 1 \end{array}$$

Change signs

$$\begin{array}{c} - \quad + \\ \underline{x - 1} \\ 0 \end{array}$$

Our answer: $x^2 + x + 1$

33. If one number is x, the other can be 4x.
The sum of the two numbers is 25.

$$\begin{aligned} x + 4x &= 25 \\ 5x &= 25 \\ x &= 5 \\ 4x &= 20 \end{aligned}$$

The two numbers are 5 and 20.

37. If we let x = the number of $5 bills, then the number of $10 bills is x + 4.

$$\begin{aligned} 5x + 10(x + 4) &= 160 \\ 5x + 10x + 40 &= 160 \\ 15x + 40 &= 160 \\ 15x &= 160 \\ x &= 8 \\ x + 4 &= 12 \end{aligned}$$

There are 8 five dollar bills and 12 ten dollar bills.

1. $(-1)^3 = (-1)(-1)(-1) = -1$

5. $\left(\dfrac{3}{7}\right)^2 = \left(\dfrac{3}{7}\right)\left(\dfrac{3}{7}\right) = \dfrac{9}{49}$

9. $x^{15} \cdot x^7 \cdot x^5 \cdot x^3 = x^{15+7+5+3} = x^{30}$

13. $(2^6)^4 = 2^{6 \cdot 4} = 2^{24}$

17. $(-2xyz)^3 = (-2)^3 x^3 y^3 z^3 = -8x^3 y^3 z^3$

21. $4x^{-5} = 4\left(\dfrac{1}{x^5}\right) = \dfrac{4}{x^5}$

25. $\dfrac{a^9}{a^{-6}} = a^{9-(-6)} = a^{15}$

29. $\dfrac{x^9}{x^{-6}} = x^{9-(-13)} = x$

33. $\dfrac{x^{-9}}{x^{-13}} = x^{-9-(-13)} = x^{-4}$

37. $(3x^3 y^2) = 3^2 (x^3)^2 (y^2)^2 = 9x^{3(2)} y^{2(2)} = 9x^6 y^4$

41. $(-3xy^2)^{-3} = \dfrac{1}{(-3xy^2)^3}$

$\qquad = \dfrac{1}{(-3)^3 x^3 (y^2)^3}$

$\qquad = \dfrac{1}{-27x^3 y^{2(3)}} \qquad (-3)^3 = (-3)(-3)(-3) = -27$

$\qquad = -\dfrac{1}{27x^3 y^6}$

45. $\dfrac{(x^{-3})^3 (x^6)^{-1}}{(x^{-5})^{-4}} = \dfrac{x^{-3(3)} x^{6(-1)}}{x^{-5(-4)}}$

$\qquad = \dfrac{x^{-9} x^{-6}}{x^{20}}$

$\qquad = \dfrac{x^{-15}}{x^{20}}$

$\qquad = \dfrac{1}{x^{20} x^{15}}$

$\qquad = \dfrac{1}{x^{35}}$

49. $\dfrac{(10x^3y^5)(21x^2y^6)}{(7xy^3)(5x^9y)}$

$$= \dfrac{10 \cdot 21x^3x^2y^5y^6}{7 \cdot 5xx^9y^3y}$$

$$= \dfrac{210x^{3+2}y^{5+6}}{35x^{1+9}y^{3+1}}$$

$$= \dfrac{6x^5y^{11}}{x^{10}y^4}$$

$$= 6x^{5-10}y^{11-4}$$

$$= 6x^{-5}y^7$$

$$= \dfrac{6y^7}{x^5}$$

53. $\dfrac{8x^8y^3}{2x^3y} - \dfrac{10x^6y^9}{5xy^7}$

$$= 4x^{8-3}y^{3-1} - 2x^{6-1}y^{9-7}$$

$$= 4x^5y^2 - 2x^5y^2$$

$$= (4 - 2)x^3y^2$$

$$= 2x^5y^2$$

57. $\dfrac{4.6 \times 10^5}{2 \times 10} = 2.3 \times 10^{5-(-3)}$

$$= 2.3 \times 10^8$$

61. $(3a^2 - 5a + 5) + (5a^2 - 7a - 8)$
$= 3a^2 + 5a^2 - 5a - 7a + 5 - 8$
$= 8a^2 - 12a - 3$

65. $(4x^2 - 3x - 2) - (8x^2 + 3x - 2)$
$= 4x^2 - 3x - 2 - 8x^2 - 3x + 2$
$= -4x^2 - 6x$

69.

x	$y = x^2 + 2$	y
-3	$y = (-3)^2 + 2 = 9 + 2 = 11$	11
-2	$y = (-2)^2 + 2 = 4 + 2 = 6$	6
-1	$y = (-1)^2 + 2 = 1 + 2 = 3$	3
0	$y = 0^2 + 2 \quad = 0 + 2 = 2$	2
1	$y = 1^2 + 2 \quad = 1 + 2 = 3$	3
2	$y = 2^2 + 2 \quad = 4 + 2 = 6$	6
3	$y = 3^2 + 2 \quad = 9 + 2 = 11$	11

vertex (at row x=0)

See the graph on page A-23 in the textbook.

73. $3x(4x - 7) = 3x(4x) + 3x(-7) = 12x^2 - 21$

77.
$$
\begin{array}{r}
a^2 + 5a - 4 \\
a + 1 \\
\hline
a^2 + 5a - 4 \\
a^3 + 5a^2 - 4a \\
\hline
a^3 + 6a^2 + a - 4
\end{array}
$$

81.
$$ \overset{F}{} \quad \overset{O}{} \quad \overset{I}{} \quad \overset{L}{}$$
$$(3x - 7)(2x - 5) = 3x(2x) + 3x(-5) + (-7)(2x) + (-7)(-5)$$
$$= 6x^2 - 15x - 14x + 35$$
$$= 6x^2 - 29x + 35$$

85. $(a^2 - 3)(a^2 + 3) = (a^2)^2 - (3)^2$
$$= a^4 - 9$$

89. $(3x + 4)^2 = (3x)^2 + 2(3x)(4) + 4^2$
$$= 9x^2 + 24x + 16$$

93. $\dfrac{10ab}{-5a} + \dfrac{20a^2}{-5a} = -2b - 4a$

97. $\dfrac{16xy^2}{-2xy} - \dfrac{10xy}{-2xy} = -8y + 5$

101.
$$
\begin{array}{r}
x + 9 \\
x + 6 \overline{)\; x^2 + 15x + 54} \\
\underline{- - } \quad \\
\underline{x^2 + 6x } \\
9x + 54 \\
\underline{- - } \\
\underline{9x + 54} \\
0
\end{array}
$$

Change signs

Change signs

The answer: $x + 9$

105.

$$
\begin{array}{r}
x^2 - 4x + 16 \\
x + 4 \overline{)\ x^3 + 0x^2 + 0x + 64}
\end{array}
$$

$$
\begin{array}{r}
\underline{-\quad -} \\
x^3 + 4x^2 \\
\hline
-4x^2 + 0x \\
+\qquad + \\
\underline{-4x^2 - 16x} \\
16x + 64 \\
-\qquad - \\
\underline{16x + 64} \\
0
\end{array}
$$

Change signs

Change signs

Change signs

The answer: $x^2 - 4x + 16$

109.

$$
\begin{array}{r}
x^2 - 4x + 5 \\
2x + 1 \overline{)\ 2x^3 - 7x^2 + 6x + 10}
\end{array}
$$

$$
\begin{array}{r}
\underline{-\quad -} \\
2x^3 + x^2 \\
\hline
-8x^2 + 6x \\
+\qquad + \\
\underline{-8x^2 - 4x} \\
10x + 10 \\
-\qquad - \\
\underline{10x + 5} \\
5
\end{array}
$$

Change signs

Change signs

Change signs

The answer: $x^2 - 4x + 5 + \dfrac{5}{2x + 1}$

Chapter 4 Test

1. $(-3)^4 = (-3)(-3)(-3)(-3)$
    $\qquad = 81$

2. $\left(\frac{3}{4}\right)^2 = \left(\frac{3}{4}\right)\left(\frac{3}{4}\right)$
    $\qquad\qquad = \frac{9}{16}$

3. $(3x^3)^2(2x^4)^3 = 3^2 \cdot x^{3\cdot2} \cdot 2^3 \cdot x^{4\cdot3}$
    $\qquad\qquad\quad = 9x^6 \cdot 8x^{12}$
    $\qquad\qquad\quad = 9 \cdot 8x^{6+12}$
    $\qquad\qquad\quad = 72x^{18}$

4. $3^{-2} = \frac{1}{3^2}$
    $\qquad = \frac{1}{9}$

5. $(3a^4b^2)\ = 1$

6. $\frac{a^{-3}}{a} = a^{-3-(-5)}$
    $\qquad = a^{-3+5}$
    $\qquad = a^2$

7. $\frac{(x^{-2})^3(x^{-3})^{-5}}{(x^{-4})^{-2}} = \frac{x^{-2\cdot3}x^{-3(-5)}}{x^{-4(-2)}}$
    $\qquad\qquad\qquad = \frac{x^{-6}x^{15}}{x^8}$
    $\qquad\qquad\qquad = \frac{x^9}{x^8}$
    $\qquad\qquad\qquad = x$

8. $0.0278 = 2.78 \times 10^{-2}$

9. $2.43 \times 10^5 = 2.43 \times 100{,}000 = 243{,}000$

10. $\dfrac{35x^2y^4z}{70x^6y^2a} = \dfrac{35}{70} \cdot \dfrac{x^2}{x^6} \cdot \dfrac{y^4}{y^2} \cdot \dfrac{z}{z}$

$\phantom{\dfrac{35x^2y^4z}{70x^6y^2a}} = \dfrac{1}{2} \cdot \dfrac{1}{x^4} \cdot \dfrac{y^2}{1} \cdot 1$

$\phantom{\dfrac{35x^2y^4z}{70x^6y^2a}} = \dfrac{y^2}{2x^4}$

11. $\dfrac{(6a^2b)(9a^3b^2)}{18a^4b^3} = \dfrac{54a^5b^3}{18a^4b^3}$

$\phantom{\dfrac{(6a^2b)(9a^3b^2)}{18a^4b^3}} = \dfrac{54}{18} \cdot \dfrac{a^5}{a^4} \cdot \dfrac{b^3}{b^3}$

$\phantom{\dfrac{(6a^2b)(9a^3b^2)}{18a^4b^3}} = \dfrac{3}{1} \cdot \dfrac{a}{1} \cdot 1$

$\phantom{\dfrac{(6a^2b)(9a^3b^2)}{18a^4b^3}} = 3a$

12. $\dfrac{24x^7}{3x^2} + \dfrac{14x^9}{7x} = 8x^5 + 2x^5$

$\phantom{\dfrac{24x^7}{3x^2} + \dfrac{14x^9}{7x}} = 10x^5$

13. $\dfrac{(2.4 \times 10^5)(4.5 \times 10^{-2})}{1.2 \times 10^{-6}} = \dfrac{(2.4)(4.5)}{1.2} \times \dfrac{10^5 \cdot 10^{-2}}{10^{-6}}$

$\phantom{\dfrac{(2.4 \times 10^5)(4.5 \times 10^{-2})}{1.2 \times 10^{-6}}} = 9.0 \times 10^9$

14.

x	$y = x^2 + 3$	y	
-3	$y = (-3)^2 + 3 = 9 + 3 = 12$	12	
-2	$y = (-2)^2 + 3 = 4 + 3 = 7$	7	
-1	$y = (-1)^2 + 3 = 1 + 3 = 4$	4	
0	$y = 0^2 + 3 \quad = 0 + 3 = 3$	3	vertex
1	$y = 1^2 + 3 \quad = 1 + 3 = 4$	4	
2	$y = 2^2 + 3 \quad = 4 + 3 = 7$	7	
3	$y = 3^2 + 3 \quad = 9 + 3 = 12$	12	

See the graph on page A-24 in the textbook.

15.

x	$y = (x + 1)^2$	y	
2	$y = (2 + 1)^2 \quad = 3^2 \quad = 9$	9	
1	$y = (1 + 1)^2 \quad = 2^2 \quad = 4$	4	
0	$y = (0 + 1)^2 \quad = 1^2 \quad = 1$	1	
-1	$y = (-1 + 1)^2 = 0^2 \quad = 0$	0	vertex
-2	$y = (-2 + 1)^2 = (-1)^2 = 1$	1	
-3	$y = (-3 + 1)^2 = (-2)^2 = 4$	4	
-4	$y = (-4 + 1)^2 = (-3)^2 = 9$	9	

See the graph on page A-24 in the textbook.

16. $8x^2 - 4x + 6x + 2 = 8x^2 + 2x + 2$

17. $(5x^2 - 3x + 4) - (2x^2 - 7x - 2)$
$= 5x^2 - 3x + 4 - 2x^2 + 7x + 2$
$= (5x^2 - 2x^2) + (-3x + 7x) + (4 + 2)$
$= 3x^2 + 4x + 6$

18. $(6x - 8) - (3x - 4) = 6x - 8 - 3x + 4$
$= (6x - 3x) + (-8 + 4)$
$= 3x - 4$

19. $2y^2 - 3y - 4$ when $y = -2$
$2(-2)^2 - 3(-2) - 4$
$= 2 \cdot 4 + 6 - 4$
$= 8 + 6 - 4$
$= 10$

20. $2a^2(3a^2 - 5a + 4) = 2a^2(3a^2) + 2a^2(-5a) + 2a^2(4)$
$= 6a^4 - 10a^3 + 8a^2$

21.

$$\overset{\textbf{F} \qquad \textbf{O} \qquad \textbf{I} \qquad \textbf{L}}{(x + \tfrac{1}{2})(x + \tfrac{1}{3}) = x(x) + x(\tfrac{1}{3}) + x(\tfrac{1}{2}) + (\tfrac{1}{2})(\tfrac{1}{3})}$$

$$= x^2 + \frac{1}{3}x + \frac{1}{2}x + \frac{1}{6}$$

$$= x^2 + \frac{5}{6}x + \frac{1}{6} \qquad\qquad \frac{1}{3} + \frac{1}{2} = \frac{2}{6} + \frac{3}{6} = \frac{5}{6}$$

22. $(4x - 5)(2x + 3) = 4x(2x) + 4x(3) + (-5)(2x) + (-5)(-3)$
$= 8x^2 + 12x - 10x - 15$
$= 8x^2 + 2x - 15$

23. $(x - 3)(x^2 + 3x + 9) = x^3 - 27$

$$
\begin{array}{r}
x^2 + 3x + \ 9 \\
x - \ \ 3 \\
\hline
- 3x^2 - 9x - 27 \\
x^3 + 3x^3 + 9x \ \ \ \ \ \ \ \ \ \ \\
\hline
x^3 \ \ \ \ \ \ \ \ \ \ \ \ \ \ \ - 27
\end{array}
$$

24. $(x + 5)^2 = (x + 5)(x + 5)$
$= x^2 + 2 \cdot 5x + 5^2$
$= x^2 + 10x + 25$

25. $(3a - 2b)^2 = (3a)^2 + 2(3a)(-2b) + (-2b)^2$
$= 9a^2 - 12ab + 4b^2$

26. $(3x - 4y)(3x + 4y) = (3x)^2 - (4y)^2$
$$= 9x^2 - 16y^2$$

27. $(a^2 - 3)(a^2 + 3) = (a^2)^2 - 3^2$
$$= a^4 - 9$$

28. $\dfrac{10x^3 + 15x^2 - 5x}{5x} = \dfrac{10x^3}{5x} + \dfrac{15x}{5x} - \dfrac{5x}{5x}$

$$= 2x^2 + 3x - 1$$

29. $\dfrac{8x^2 - 6x - 5}{2x - 3}$

$$
\begin{array}{r}
4x + 3 \\
2x - 3 \overline{) 8x^2 - 6x - 5} \\
 - \quad + \\
 \underline{8x^2 - 12x} \\
6x - 5 \\
- \quad + \\
\underline{6x - 9} \\
4
\end{array}
$$

Change signs

Change signs

Our answer is: $4x + 3 + \dfrac{4}{2x - 3}$

30. $\dfrac{3x^3 - 2x + 1}{x - 3}$

$$
\begin{array}{r}
3x^2 + 9x + 25 \\
x - 3 \overline{) 3x^3 + 0x^2 - 2x + 1} \\
- \quad + \\
\underline{3x^3 - 9x^2} \\
9x^2 - 2x \\
- \quad + \\
\underline{9x^2 - 27x} \\
25x + 1 \\
- \quad + \\
\underline{25x - 75} \\
76
\end{array}
$$

Our answer is: $3x^2 + 9x + 25 + \dfrac{76}{x - 3}$

Chapter 5

Section 5.1

1. $15x + 25 = 5 \cdot 3x + 5 \cdot 5$
$\qquad\qquad = 5(3x + 5)$

5. $4x - 8y = 4(x) - 4(2y)$
$\qquad\qquad = 4(x - 2y)$

9. $3a^2 - 3a - 60 = 3(a^2) - 3(a) - 3(20)$
$\qquad\qquad\qquad = 3(a^2 - a - 20)$

13. $9x^2 - 8x^3 = x^2(9) - x^2(8x)$
$\qquad\qquad\quad = x^2(9 - 8x)$

17. $21x^2y - 28xy^2 = 7xy(3x) - 7xy(4y)$
$\qquad\qquad\qquad = 7xy(3x - 4y)$

21. $7x^3 + 21x^2 - 28x = 7x(x^2) + 7x(3x) - 7x(4)$
$\qquad\qquad\qquad\quad = 7x(x^2 + 3x - 4)$

25. $100x^4 - 50x^3 + 25x^2 = 25x^2(4x^2) - 25x^2(2x) + 25x^2(1)$
$\qquad\qquad\qquad\qquad = 25x^2(4x^2 - 2x + 1)$

29. $4a^2b - 16ab^2 + 32a^2b^2 = 4ab(a) - 4ab(4b) + 4ab(8ab)$
$\qquad\qquad\qquad\qquad = 4ab(a - 4b + 8ab)$

33. $12x^2y^3 - 72x^5y^3 - 36x^4y^4 = 12x^2y^3(1) - 12x^2y^3(6x^3) - 12x^2y^3(3x^2y)$
$\qquad\qquad\qquad\qquad\qquad = 12x^2y^3(1 - 6x^3 - 3x^2y)$

37. $xy + 6x + 2y + 12 = x(y + 6) + 2(y + 6)$
$\qquad\qquad\qquad = (y + 6)(x + 2)$

41. $ax - bx + ay - by = x(a - b) + y(a - b)$
$\qquad\qquad\qquad = (a - b)(x + y)$

45. $3xb - 4b - 6x + 8 = b(3x - 4) - 2(3x - 4)$
$\qquad\qquad\qquad = (3x - 4)(b - 2)$

49. $x^2 - ax - bx + ab = x(x - a) - b(x - a)$
$\qquad\qquad\qquad = (x - a)(x - b)$

53. $x^3 + 2x^2 + 3x + 6 = x^2(x + 2) + 3(x + 2)$
$\qquad\qquad\qquad = (x^2 + 3)(x + 2)$

57. $(3x + 6)(2x + 4) = 6x^2 + 24x + 24$
$\qquad\qquad\qquad = 6(x^2 + 4x + 4)$

The greatest common factor is 6.

Section 5.1 continued

61. $A = 1,000 + 1,000r$

$\quad = 1,000(1 + r)$

When $\qquad\qquad r = .12$
the equation $\quad A = 1,000(1 + r)$
becomes $\qquad\quad A = 1,000(1 + .12)$
$\qquad\qquad\qquad\quad = 1,000(1.12)$
$\qquad\qquad\qquad\quad = \$1,120$

65. $(x + 7)(x - 2) = x \cdot x + x(-2) + 7(x) + 7(-2)$

$\qquad\qquad\qquad\quad$ F $\qquad$ O $\qquad$ I $\qquad$ L

$\qquad\qquad = \quad x^2 \quad - \quad 2x \; + \; 7x \; - \; 14$

$\qquad\qquad = x^2 + 5x - 14$

69. $(x - 3)(x + 2) = x(x) + x(2) - 3(x) - 3(2)$

$\qquad\qquad\qquad\quad$ F $\qquad$ O $\qquad$ I $\qquad$ L

$\qquad\qquad = \quad x^2 \; + \; 2x \; - \; 3x \; - \; 6$

$\qquad\qquad = x^2 - x - 6$

73.
$$
\begin{array}{r}
x^2 + 4x - 3 \\
2x + 1 \\
\hline
x^2 + 4x - 3 \\
3x^3 + 8x^2 - 6x \phantom{{}-3} \\
\hline
3x^3 + 9x^2 - 2x - 3
\end{array}
$$

Section 5.2

1. We need to find a pair of numbers whose sum is 7 and whose product is 12.

$\qquad x^2 + 7x + 12 = (x + 3)(x + 4)$

5. We need to find a pair of numbers whose sum is 10 and whose product is 21.

$\qquad a^2 + 10a + 21 = (a + 3)(a + 7)$

9. We need to find a pair of numbers whose sum is -10 and whose product is 21.

$\qquad y^2 - 10y + 21 = (y - 3)(y - 7)$

13. We need to find a pair of numbers whose difference is 1 and whose product is -12.

$\qquad y^2 + y - 12 = (y + 4)(y - 3)$

17. We need to find a pair of numbers whose difference is -8 and whose product is -9.

$\qquad r^2 - 8r - 9 = (r - 9)(r + 1)$

133

Section 5.2 continued

21. We need to find a pair of numbers whose sum is 15 and whose product is 56.

$$a^2 + 15a + 56 = (a + 7)(a + 8)$$

25. We need to find a pair of numbers whose sum is 13 and whose product is 42.

$$x^2 + 13x + 42 = (x + 6)(x + 7)$$

29. $3a^2 - 3a - 60 = 3(a^2) - 3(a) - 3(20)$
$$= 3(a^2 - a - 20)$$

We need to find a pair of numbers whose difference is -1 and whose product is -20.

$$3(a^2 - a - 20) = 3(a + 4)(a - 5)$$

33. $100p^2 - 1300p + 4000 = 100(p^2) - 100(13p) + 100(40)$
$$= 100(p^2 - 13p + 40)$$

We need to find a pair of numbers whose sum is -13 and whose product is 40.

$$100(p^2 - 13p + 40) = 100(p - 5)(p - 8)$$

37. $2r^3 + 4r^2 - 30r = 2r(r^2) + 2r(2r) - 2r(15)$
$$= 2r(r^2 + 2r - 15)$$

We need to find a pair of numbers whose difference is 2 and whose product is -15.

$$2r(r^2 + 2r - 15) = 2r(r + 5)(r - 3)$$

41. $x^5 + 4x^4 + 4x^3 = x^3(x^2) + x^2(4x) + x^3(4)$
$$= x^3(x^2 + 4x + 4)$$

We need to find a pair of numbers whose sum is 4 and whose product is 4.

$$x^3(x^2 + 4x + 4) = x^3(x + 2)(x + 2)$$

45. $4x^4 - 52x^3 + 144x^2 = 4x^2(x^2) - 4x^2(13x) + 4x^2(36)$
$$= 4x^2(x^2 - 13x + 36)$$

We need to find a pair of numbers whose sum is -13 and whose product is 36.

$$4x (x - 13x + 36) = 4x (x - 4)(x - 9)$$

49. We need to find a pair of numbers whose sum is -9 and whose product is 20.

$$x^2 - 9xy + 20y^2 = (x - 4y)(x - 5y)$$

Section 5.2 continued

53. We need to find a pair of numbers whose sum is -10 and whose product is 25.

$$a^2 - 10ab + 25b^2 = (a - 5b)(a - 5b)$$

57. We need to find a pair of numbers whose sum is 2 and whose product is -48.

$$x^2 + 2xa - 48a^2 = (x - 6a)(x + 8a)$$

61. $x^2 - 5x^2 + 6 = (x^2 - 6)(x^2 + 1)$

65. $x^2 - x + \frac{1}{4} = (x - \frac{1}{2})(x - \frac{1}{2})$ $\frac{1}{2} \cdot \frac{1}{2} = \frac{1}{4}$

69. $x^2 + 24x + 128 = (x + 8)(x + 16)$

The other factor is $x + 16$.

73.
$$\begin{aligned} h &= 64 + 48t - 16t^2 \\ &= 16(4 + 3t - t^2) \\ &= 16(4 - t)(1 + t) \end{aligned}$$

When   t = 4
the equation $h = 16(4 - t)(1 + t)$
becomes $h = 16(4 - 4)(1 + 4)$
                         $= 16(0)(5)$
                         $= 0$

When the time is 4 seconds, the height of the arrow is 0.

77. $(3a + 2)(2a + 1) = 3a(2a) + 3a(1) + 2(2a) + 2(1)$

                                    F   0   I   L

$$\begin{aligned} &= 6a^2 \quad + 3a \quad + 4a \quad + 2 \\ &= 6a^2 + 7a + 2 \end{aligned}$$

81. $(5x^2 + 5x - 4) - (3x^2 - 2x + 7)$
$$\begin{aligned} &= 5x^2 + 5x - 4 - 3x^2 + 2x - 7 \qquad \text{Subtract} \\ &= 2x^2 + 7x - 11 \end{aligned}$$

85. $(5x^2 - 5) - (2x^2 - 4x)$
$$\begin{aligned} &= 5x^2 - 5 - 2x^2 + 4x \qquad \text{Subtract} \\ &= 3x^2 + 4x - 5 \end{aligned}$$

Section 5.3

1. Factor: $2x^2 + 7x + 3$

   Possible Factors   Middle Term when Multiplied
   $(2x + 1)(x + 3)$ $7x$
   $(2x + 3)(x + 1)$ $5x$

   $$2x^2 + 7x + 3 = (2x + 1)(x + 3)$$

5. Factor: $3x^2 + 2x - 5$

   Possible Factors   Middle Term when Multiplied
   $(3x + 1)(x - 5)$ $-14x$
   $(3x + 5)(x - 1)$ $2x$
   $(x + 1)(3x - 5)$ $-2x$
   $(x + 5)(3x - 1)$ $14x$

   $$3x^2 + 2x - 5 = (3x + 5)(x - 1)$$

9. Factor: $6x^2 + 13x + 6$

   Possible Factors   Middle Term when Multiplied
   $(6x + 1)(x + 6)$ $37x$
   $(6x + 6)(x + 1)$ $12x$
   $(6x + 2)(x + 3)$ $20x$
   $(6x + 3)(x + 2)$ $15x$
   $(2x + 3)(3x + 2)$ $13x$
   $(2x + 2)(3x + 3)$ $12x$

   $$6x^2 + 13x + 6 = (2x + 3)(3x + 2)$$

13. Factor: $4y^2 - 11y - 3$

   Possible Factors   Middle Term when Multiplied
   $(4y + 1)(y - 3)$ $-11y$
   $(4y - 3)(y + 1)$ y
   $(2y + 1)(2y - 3)$ $-4y$
   $(2y - 1)(2y + 3)$ $4y$

   $$4y^2 - 11y - 3 = (4y + 1)(y - 3)$$

17. Factor: $20a^2 + 48ab - 5b^2$

   Possible Factors   Middle Term when Multiplied
   $(20a + 1b)(a - 50)$ $-99ab$
   $(20a - 5b)(a + 1b)$ $15ab$
   $(4a + 1b)(5a - 5b)$ $-15ab$
   $(4a - 5b)(5a + 1b)$ $-21ab$
   $(2a + 1b)(10a - 5b)$ $0ab$
   $(2a + 5b)(10a - 1b)$ $48ab$

   $$20a^2 + 48ab - 5b^2 = (2a + 5b)(10a - b)$$

21. Factor: $12m^2 + 16m - 3$

Possible Factors	Middle Term when Multiplied
$(12m + 1)(m - 3)$	$-35m$
$(12m - 3)(m + 1)$	$9m$
$(3m + 1)(4m - 3)$	$-5m$
$(3m - 3)(4m + 1)$	$-9m$
$(2m + 1)(6m - 3)$	$0m$
$(2m - 3)(6m + 1)$	$-16m$
$(2m + 1)(6m - 3)$	$0m$
$(2m + 3)(6m - 1)$	$16m$
$(2m - 1)(6m + 3)$	$0m$

$$12m^2 + 16m - 3 = (2m + 3)(6m - 1)$$

25. Factor: $12a^2 - 25ab + 12b^2$

Possible Factors	Middle Term when Multiplied
$(12a - 1b)(a - 12b)$	$-145ab$
$(12a - 12)(a - 1b)$	$-24ab$
$(3a - 1b)(4a - 12b)$	$-40ab$
$(3a - 12b)(4a - 1b)$	$-51ab$
$(2a - 1b)(6a - 12b)$	$-30ab$
$(2a - 12b)(6a - 1b)$	$-74ab$
$(2a - 3b)(6a - 4b)$	$-26ab$
$(2a - 4b)(6a - 3b)$	$-30ab$
$(3a - 3b)(4a - 4b)$	$-24ab$
$(3a - 4b)(4a - 3b)$	$-25ab$

$$12a^2 - 25ab + 12b^2 = (3a - 4b)(4a - 3b)$$

29. Factor: $14x^2 + 29x - 15$

Possible Factors	Middle Term when Multiplied
$(2x - 5)(7x + 3)$	$-29x$
$(2x + 5)(7x - 3)$	$29x$
$(14x - 5)(1x + 3)$	$37x$
$(14x + 5)(1x - 3)$	$-37x$

$$14x^2 + 29x - 15 = (2x + 5)(7x - 3)$$

33. Factor: $15t^2 - 67t + 38$

Possible Factors	Middle Term when Multiplied
$(3t - 2)(5t - 19)$	$-67t$
$(3t + 2)(5t + 19)$	$67t$
$(5t - 2)(5t - 19)$	$-105t$
$(5t + 2)(5t + 19)$	$105t$

$$15t^2 - 67t + 38 = (3t - 2)(5t - 19)$$

37. Factor: $24a^2 - 50a + 24 = 2(12a^2 - 25a + 12)$

See possible factors in #25.

$$2(12a^2 - 25a + 12) = 2(3a - 4)(4a - 3)$$

41. Factor: $6x^4 - 11x^3 - 10x^2 = x^2(6x^2 - 11x - 10)$

Possible Factors	Middle Term when Multiplied
$(6x - 2)(x + 5)$	$28x$
$(6x + 5)(x - 2)$	$- 7x$
$(3x - 2)(2x + 5)$	$11x$
$(3x + 2)(2x - 5)$	$-11x$

$$x^2(6x^2 - 11x - 10) = x^2(3x + 2)(3x - 5)$$

45. Factor: $15x^3 - 102x^2 - 21x = 3x(5x^2 - 34x - 7)$

Possible Factors	Middle Term when Multiplied
$(5x + 1)(x - 7)$	$-34x$
$(5x - 7)(x - 7)$	$- 2x$

$$3x(5x^2 - 34x - 7) = 3x(5x + 1)(x - 7)$$

49. Factor: $15a^4 - 2a^3 - a^2 = a^2(15a^2 - 2a - 1)$

Possible Factors	Middle Term when Multiplied
$(5a - 1)(3a + 1)$	$2a$
$(5a + 1)(3a - 1)$	$- 2a$
$(1x - 1)(15x + 1)$	$-14a$
$(1x + 1)(15x - 1)$	$14a$

$$a^2(15a^2 - 2a - 1) = a^2(5a + 1)(3a - 1)$$

53. Factor: $12x^2y - 34xy^2 + 14y^3 = 2y(6x^2 - 17xy + 7y^2)$

Possible Factors	Middle Term when Multiplied
$(3x - 1)(2x - 7y)$	$-23xy$
$(3x - 7)(2x - 1y)$	$-17xy$
$(6x - 1)(1x - 7y)$	$-43xy$
$(6x - 7)(1x - 1y)$	$-13xy$

$$2y(6x^2 - 17xy + 7xy^2) = 2y(3x - 7y)(2x - y)$$

57. $(2x + 3)(2x - 3) = 4x^2 - 6x + 6x - 9$
$$= 4x^2 - 9$$

61. $(x + 3)(x - 3) = x^2 - 3x + 3x - 9$
$$= x^2 - 9$$

65. $(x + 4)^2 = (x + 4)(x + 4)$
$$= x^2 + 4x + 4x + 16$$
$$= x^2 + 8x + 16$$

69. $\dfrac{12x^6 - 18x^4 + 24x^3}{6x} = \dfrac{12x^6}{6x^2} - \dfrac{18x^4}{6x^2} + \dfrac{24x^3}{6x^2}$

$$= 2x^4 - 3x^2 + 4x$$

Section 5.4

1. $x^2 - 9 = (x)^2 - (3)^2$
   $ = (x + 3)(x - 3)$

5. $x^2 - 49 = (x)^2 - (7)^2$
   $ = (x + 7)(x - 7)$

9. $9x^2 + 25 \qquad$ No factors

13. $9x^2 - 16b^2 = (3x)^2 - (4b)^2$
    $ = (3x + 4b)(3x - 4b)$

17. $25 - 4x^2 = (5)^2 - (2x)^2$
    $ = (5 + 2x)(5 - 2x)$

21. $32a^2 - 128 = 32(a^2 - 4)$
    $ = 32(a^2 - 2^2)$
    $ = 32(a + 2)(a - 2)$

25. $a^4 - b^4 = (a^2)^2 - (b^2)^2$
    $ = (a^2 + b^2)(a^2 - b^2)$
    $ = (a^2 + b^2)(a + b)(a - b)$

29. $3x^3y - 75xy^3 = 3xy(x^2 - 25y^2)$
    $ = 3xy[x^2 - (5y)^2]$
    $ = 3xy(x + 5y)(x - 5y)$

33. $x^2 + 2x + 1 = (x + 1)(x + 1)$
    $ = (x + 1)^2$

37. $y^2 + 4y + 4 = (y + 2)(y + 2)$
    $ = (y + 2)^2$

41. $m^2 - 12m + 36 = (m - 6)(m - 6)$
    $ = (m - 6)^2$

45. $49x^2 - 14x + 1 = (7x - 1)(7x - 1)$
    $ = (7x - 1)^2$

49. $x^2 + 10xy + 25y^2 = (x + 5y)(x + 5y)$
    $ = (x + 5y)^2$

53. $3a^2 + 18a + 27 = 3(a^2 + 6a + 9)$
    $ = 3(a + 3)(a + 3)$
    $ = 3(a + 3)^2$

57. $5x^3 + 30x^2y + 45xy^2 = 5x(x^2 + 6xy + 9y^2)$
    $ = 5x(x + 3y)(x + 3y)$
    $ = 5x(x + 3y)^2$

61. $x^3 + 3x^2 - 4x - 12$

$= (x^3 + 3x^2) + (-4x - 12)$ Group

$= x^2(x + 3) - 4(x + 3)$ Factor

$= (x + 3)(x^2 - 4)$

$= (x + 3)(x + 2)(x - 2)$

65. $x^2 + 6x + 9 - y^2$

$= (x^2 + 6x + 9) - y^2$ Group first 3 terms together

$= (x + 3)^2 - y^2$ This has the form $a^2 - b^2$

$= [(x + 3) + y][(x + 3) - y]$ Factoring according to the formula $a^2 - b^2 = (a + b)(a - b)$

$= (x + 3 + y)(x + 3 - y)$ Simplify

69. $x^2 + bx + 49 = (x + 7)^2$

$x^2 + 14x + 49 = (x + 7)^2$

$b = 14$

73.

$$x - 3 \overline{\smash{)}\ x^2 - 5x + 3} \qquad x - 2 + \frac{2}{x - 3}$$

$$\underline{\quad -\qquad +\qquad}$$
$$\underline{x^2 - 3x}$$
$$- 2x + 8$$
$$\underline{\quad +\qquad -\qquad}$$
$$\underline{- 2x + 6}$$
$$2$$

Change signs

Change signs

77.

$$2x + 3 \overline{\smash{)}\ 6x^2 + 5x + 3} \qquad 3x - 2 + \frac{9}{2x + 3}$$

$$\underline{\quad -\qquad -\qquad}$$
$$\underline{6x^2 + 9x}$$
$$- 4x + 3$$
$$\underline{\quad +\qquad +\qquad}$$
$$\underline{- 4x - 6}$$
$$9$$

Change signs

Change signs

Section 5.5

1. $x^2 - 81 = (x + 9)(x - 9)$

5. $x^2 + 6x + 9 = (x + 3)(x + 3)$

$= (x + 3)^2$

9. $2a^3b + 6a^2b + 2ab = 2ab(a^2 + 3a + 1)$

13. $12a^2 - 75 = 3(4a^2 - 25)$

$= 3(2a + 5)(2a - 5)$

17. $4x^3 + 16xy^2 = 4x(x^2 + 4y^2)$

21. $a^6 + 4a^4b^2 = a^4(a^2 + 4b^2)$

25. $x^4 - 16 = (x^2 + 4)(x^2 - 4)$
$$= (x^2 + 4)(x + 2)(x - 2)$$

29. $5a^2 + 10ab + 5b^2 = 5(a^2 + 2ab + b^2)$
$$= 5(a + b)(a + b)$$
$$= 5(a + b)^2$$

33. $3x^2 + 15xy + 18y^2 = 3(x^2 + 5xy + 6y^2)$
$$= 3(x + 2y)(x + 3y)$$

37. $100x^2 - 300x + 200 = 100(x^2 - 3x + 2)$
$$= 100(x - 2)(x - 1)$$

41. $x^2 + 3x + ax + 3a = (x^2 + 3x) + (ax + 3a)$
$$= x(x + 3) + a(x + 3)$$
$$= (x + 3)(x + a)$$

45. $49x^2 + 9y^2$ Cannot be factored.

49. $xa - xb + ay - by = (xa - xb) + (ay - by)$
$$= x(a - b) + y(a - b)$$
$$= (a - b)(x + y)$$

53. $20x^4 - 45x^2 = 5x^2(4x^2 - 9)$
$$= 5x^2(2x + 3)(2x - 3)$$

57. $16x^5 - 44x^4 + 30x^3 = 2x^3(8x^2 - 22x + 15)$
$$= 2x^3(4x - 5)(2x - 3)$$

61. $y^4 - 1 = (y^2 + 1)(y^2 - 1)$
$$= (y^2 + 1)(y + 1)(y - 1)$$

65. $3x - 6 = 9$
$$3x = 15$$
$$x = 5$$

69. $4x + 3 = 0$
$$4x = -3$$
$$x = -\frac{3}{4}$$

Section 5.5 continued

73. $(3x^3)^2(2x^4)^3 = 3^2 \cdot x^{3(2)}2^3 \cdot x^{4(3)}$

$\qquad\qquad\qquad = 9x^6 \cdot 8x^{12}$

$\qquad\qquad\qquad = 7x^{6+12}$

$\qquad\qquad\qquad = 72x^{18}$

Section 5.6

1. $(x + 2)(x - 1) = 0$

$x + 2 = 0 \quad$ or $\quad x - 1 = 0 \qquad$ Set factors to 0
$\qquad x = -2 \qquad\qquad x = 1 \qquad$ Solutions

5. $x(x + 1)(x - 3) = 0$

$x = 0 \quad$ or $\quad x + 1 = 0 \quad$ or $\quad x - 3 = 0 \qquad$ Set factors to 0
$x = 0 \qquad\qquad x = -1 \qquad\qquad x = 3 \qquad$ Solutions

9. $m(3m + 4)(3m - 4) = 0$

$m = 0 \quad$ or $\quad 3m + 4 = 0 \quad$ or $\quad 3m - 4 = 0 \qquad$ Set factors to 0
$m = 0 \quad$ or $\qquad 3m = -4 \quad$ or $\qquad 3m = 4$

$m = 0 \qquad\qquad m = \dfrac{-4}{3} \qquad\quad m = \dfrac{4}{3} \qquad$ Solutions

13. $x^2 + 3x + 2 = 0$
$\qquad (x + 2)(x + 1) = 0 \qquad$ Factor the left side

$x + 2 = 0 \quad$ or $\quad x + 1 = 0 \qquad$ Set factors to 0
$\qquad x = -2 \qquad\qquad x = -1 \qquad$ Solutions

17. $a^2 - 2a - 24 = 0$
$\qquad (x + 4)(x - 6) = 0 \qquad$ Factor the left side

$x + 4 = 0 \quad$ or $\quad x - 6 = 0 \qquad$ Set factors to 0
$\qquad x = -4 \qquad\qquad x = 6 \qquad$ Solutions

21. $x^2 = -6x - 9$
$\qquad x^2 + 6x + 9 = 0 \qquad$ Standard form
$\qquad (x + 3)(x + 3) = 0 \qquad$ Factor the left side

$x + 3 = 0 \quad$ or $\quad x + 3 = 0 \qquad$ Set factors to 0
$\qquad x = -3 \qquad\qquad x = -3 \qquad$ Solutions

25. $2x^2 + 5x - 12 = 0$
$\qquad (x + 4)(2x - 3) = 0 \qquad$ Factor the left side

$x + 4 = 0 \quad$ or $\quad 2x - 3 = 0 \qquad$ Set factors to 0
$\qquad x = -4 \quad$ or $\qquad 2x = 3$

$\qquad x = -4 \qquad\qquad x = \dfrac{3}{2} \qquad$ Solutions

29.
$$a^2 + 25 = 10a$$
$$a^2 - 10a + 25 = 0 \quad \text{Standard form}$$
$$(a - 5)(a - 5) = 0 \quad \text{Factor left side}$$

$$a - 5 = 0 \quad \text{or} \quad a - 5 = 0 \quad \text{Set factors to 0}$$
$$a = 5 \qquad\qquad a = 5 \quad \text{Solutions}$$

33.
$$3m^2 = 20 - 7m$$
$$3m^2 + 7m - 20 = 0 \quad \text{Standard form}$$
$$(m + 4)(3m - 5) = 0 \quad \text{Factor left side}$$

$$m + 4 = 0 \quad \text{or} \quad 3m - 5 = 0 \quad \text{Set factors to 0}$$
$$m = -4 \quad \text{or} \qquad 3m = 5$$

$$m = -4 \qquad\qquad m = \frac{5}{3} \quad \text{Solutions}$$

37.
$$x^2 + 6x = 0$$
$$x(x + 6) = 0$$

$$x = 0 \quad \text{or} \quad x + 6 = 0$$
$$x = 0 \qquad\qquad x = -6$$

41.
$$2x^2 = 8x$$
$$2x^2 - 8x = 0$$
$$2x(x - 4) = 0$$

$$2x = 0 \quad \text{or} \quad x - 4 = 0$$
$$x = 0 \qquad\qquad x = 4$$

45.
$$1400 = 400 + 700x - 100x^2$$
$$100x^2 - 700x + 1000 = 0$$
$$100(x^2 - 7x + 10) = 0$$
$$100(x - 2)(x - 5) = 0$$

$$x - 2 = 0 \quad \text{or} \quad x - 5 = 0$$
$$x = 2 \qquad\qquad x = 5$$

49.
$$x(2x - 3) = 20$$
$$2x^2 - 3x = 20$$
$$2x^2 - 3x - 20 = 0$$
$$(2x + 5)(x - 4) = 0$$

$$2x + 5 = 0 \quad \text{or} \quad x - 4 = 0$$
$$2x = -5 \quad \text{or} \quad x = 4$$

$$x - \frac{-5}{2} \qquad\qquad x = 4$$

53.
$$4000 = (1300 - 100p)p$$
$$4000 = 1300p - 100p^2$$
$$100p^2 - 1300 + 4000 = 0$$
$$100(p^2 - 13p + 40) = 0$$
$$100(p - 5)(p - 8) = 0$$

$$p - 5 = 0 \quad \text{or} \quad p - 8 = 0$$
$$p = 5 \qquad\qquad p = 8$$

57.
$$(x + 5)^2 = 2x + 9$$
$$x^2 + 10x + 25 = 2x + 9$$
$$x^2 + 8x + 16 = 0$$
$$(x + 4)(x + 4) = 0$$

$x + 4 = 0$ or $x + 4 = 0$
$x = -4$ $x = -4$

61.
$$10^2 = (x + 2)^2 + x^2$$
$$100 = x^2 + 4x + 4 + x^2$$
$$0 = 2x^2 + 4x - 96$$
$$2(x^2 + 2x - 48) = 0$$
$$2(x + 8)(x - 6) = 0$$

$8 + x = 0$ or $x - 6 = 0$
$x = -8$ $x = 6$

65.
$$4y^3 - 2y^2 - 30y = 0$$
$$y(4y^2 - 2y - 30) = 0$$
$$y(4y + 10)(y - 3) = 0$$

$y = 0$ or $4y + 10 = 0$ or $y - 3 = 0$
$y = 0$ or $4y = -10$ or $y = 3$

$y = 0$ $y = -\dfrac{5}{2}$ $y = 3$

69.
$$20a^3 = -18a^2 + 18a$$
$$20a^3 + 18a^2 - 18a = 0$$
$$2a(10a^2 + 9a - 9) = 0$$
$$2a(2a + 3)(5a - 3) = 0$$

$2a = 0$ or $2a + 3 = 0$ or $5a - 3 = 0$
$a = 0$ $2a = -3$ or $5a = 3$

$a = 0$ $a = \dfrac{-3}{2}$ $a = \dfrac{3}{5}$

73.
$$x^3 + x^2 - 16x - 16 = 0$$
$$x^2(x + 1) - 16(x + 1) = 0$$
$$(x^2 - 16)(x + 1) = 0$$
$$(x + 4)(x - 4)(x + 1) = 0$$

$x + 4 = 0$ or $x - 4 = 0$ or $x + 1 = 0$
$x = -4$ or $x = 4$ or $x = -1$

77.   lot = x
house = $4x$
A house and a lot cost $3000.

$4x$ + x $= 3000$
                    $5x = 3000$
                    $x = \$600$ (lot)
                    $4x = \$2400$ (house)

81. $\dfrac{x^5}{x^{-3}} = x^{5-(-3)} = x^8$

85. $0.0056 = 5.6 \times 10^{-3}$

Section 5.7

1. Let x = the first even integer, then $x + 2$ = the second even integer.
$$x(x + 2) = 80$$
$$x^2 + 2x = 80$$
$$x^2 + 2x - 80 = 0$$
$$(x + 10)(x - 8) = 0$$

$x + 10 = 0 \quad$ or $\quad x - 8 = 0$

$\qquad x = -10 \quad$ or $\qquad x = 8$

$x + 2 = -2 \qquad\qquad x + 2 = 10$

The solutions are $\{-10, -8\}$ and $\{8, 10\}$.

5. Let x = the first even integer, then $x + 2$ = the second even integer.
$$x(x + 2) = 5(x + x + 2) - 10$$
$$x^2 + 2x = 5(2x + 2) - 10$$
$$x^2 + 2x = 10x + 10 - 10$$
$$x^2 - 8x = 0$$
$$x(x - 8) = 0$$

$\quad x = 0 \quad$ or $\quad x - 8 = 0$

$x + 2 = 2 \qquad\qquad x = 8$

$\qquad\qquad\qquad x + 2 = 10$

The solutions are $\{0, 2\}$ and $\{8, 10\}$.

9. Let one number = x, then the other number is $5x + 2$.
$$x(5x + 2) = 24$$
$$5x^2 + 2x = 24$$
$$5x^2 + 2x - 24 = 0$$
$$(5x + 12)(x - 2) = 0$$

$5x + 12 = 0 \quad$ or $\quad x - 2 = 0$

$\quad 5x = -12 \quad$ or $\qquad x = 2$

$\quad x = \dfrac{-12}{5} \qquad 5x + 2 = 12$

$5x + 2 = -10$

The solutions are $\{\dfrac{-12}{5}, -10\}$ and $\{2, 12\}$.

13. Let the width = x, then the length = x + 1
Area = Length · Width

$$12 = (x + 1) \cdot x$$
$$12 = x^2 + x$$
$$0 = x^2 + x - 12$$
$$0 = (x + 4)(x - 3)$$

$$0 = x + 4 \quad \text{or} \quad 0 = x - 3$$
$$-4 = x \quad \quad \text{or} \quad 3 = x$$

The solution x = -4 cannot be used, since the length and width are always given in positive units.

Width = 3 inches
Length = 4 inches

17. Let x = the shortest side, then x + 2 = the middle side.

$$10^2 = x^2 + (x + 2)^2$$
$$100 = x^2 + x^2 + 4x + 4$$
$$0 = 2x^2 + 4x - 96$$
$$0 = 2(x^2 + 2x - 48)$$
$$0 = 2(x + 8)(x - 6)$$

$$0 = x + 8 \quad \text{or} \quad 0 = x - 6$$
$$-8 = x \quad \quad \text{or} \quad 6 = x$$

The solution x = -8 cannot be used, since the length is always given in positive units.

The shorter two sides of the triangle are 6 inches and 8 inches.

21. When C = 1400, C = 400 + 700x - 100x² becomes

$$1400 = 400 + 700x - 100x^2$$
$$100x^2 - 700x + 1000 = 0$$
$$100(x^2 - 7x + 10) = 0$$
$$100(x - 2)(x - 5) = 0$$

$$x - 2 = 0 \quad \text{or} \quad x - 5 = 0$$
$$x = 2 \quad \text{or} \quad x = 5$$

The solutions are 2 hundred items and 5 hundred items.

25. When R = 3,200, R = xp = (1200 - 100p)p becomes

$$(1200 - 100p)p = 3200$$
$$1200p - 100p^2 = 3200$$
$$0 = 100p^2 - 1200p + 3200$$
$$0 = 100(p^2 - 12p + 32)$$
$$0 = 100(p - 8)(p - 4)$$

$$0 = p - 8 \quad \text{or} \quad 0 = p - 4$$
$$8 = p \quad \quad \text{or} \quad 4 = p$$

The solutions are $8 and $4 for the price of the ribbons.

29. $(6a^2b)(7a^3b^2) = 42a^{2+3}b^{1+2}$

$\qquad\qquad\qquad\qquad = 42a^5b^3$

33. $\dfrac{(5x^4y^4)(10x^3y^3)}{2x^2y^7} = \dfrac{50x^{4+3}\ y^{4+3}}{2x^2y^7}$

$\qquad\qquad\qquad = \dfrac{50x^7y^7}{2x^2y^7}$

$\qquad\qquad\qquad = 25x^{7-2}y^{7-7}$

$\qquad\qquad\qquad = 25x^5$

37. $\dfrac{45a^6}{9a^3} - \dfrac{15a^8}{5a^5} = \dfrac{225a^8}{45a^5} - \dfrac{135a^8}{45a^5}$

$\qquad\qquad\qquad = \dfrac{90a^8}{45a^5}$

$\qquad\qquad\qquad = 2a^{8-5}$

$\qquad\qquad\qquad = 2a^3$

Chapter 5 Review

1. $10x - 20 = 10 \cdot x - 10 \cdot 2$
$\qquad\qquad = 10(x - 2)$

5. $8x + 4 = 4 \cdot 2x + 4 \cdot 1$
$\qquad\quad = 4(2x + 1)$

9. $30xy^3 - 45x^3y^2 = 15xy^2 \cdot 2y - 15xy^2 \cdot 3x^2$
$\qquad\qquad\qquad\quad = 15xy^2(2y - 3x^2)$

13. $6ab^2 + 18a^3b^3 - 24a^2b = 6ab \cdot b + 6ab \cdot 3a^2b^2 - 6ab \cdot 4a$
$\qquad\qquad\qquad\qquad\quad = 6ab(b + 3a^2b^2 - 4a)$

17. $xy + 4x - 5y - 20 = x(y + 4) - 5(y + 4)$
$\qquad\qquad\qquad\qquad = (y + 4)(x - 5)$

21. $5x^2 - 4ax - 10bx + 8ab = x(5x - 4a) - 2b(5x - 4a)$
$\qquad\qquad\qquad\qquad\qquad = (5x - 4a)(x - 2b)$

25. $a^2 - 14a + 48$ $\qquad$ The coefficient of a^2 is 1.

We need two numbers whose sum is -14 and whose product is 48. The numbers are -6 and -8.

$\qquad a^2 - 14a + 48 = (a - 6)(a - 8)$

29. $y^2 + 20y + 99$ $\qquad$ The coefficient of y^2 is 1.

We need two numbers whose sum is 20 and whose product is 99. The numbers are 9 and 11.

$\qquad y^2 + 20y + 99 = (y + 9)(y + 11)$

33. $2x^2 + 13x + 15$

Possible Factors	Middle Term When Multiplied
$(2x + 3)(x + 5)$	$13x$
$(2x + 5)(x + 3)$	$11x$
$(2x + 1)(x + 15)$	$31x$
$(2x + 5)(x + 1)$	$17x$

The complete problem is

$\qquad 2x^2 + 13x + 15 = (2x + 3)(x + 5)$

37. $20a^2 - 27a + 9$

Possible Factors	Middle Term When Multiplied
$(x - 3)(20x - 3)$	$- 63x$
$(4x - 3)(5x - 3)$	$- 27x$
$(2x - 3)(10x - 3)$	$- 26x$
$(x - 1)(20x - 9)$	$- 29x$
$(20x - 1)(x - 9)$	$-181x$
$(4x - 1)(5x - 9)$	$- 41x$
$(5x - 1)(4x - 9)$	$- 49x$
$(2x - 1)(10x - 9)$	$- 28x$
$(10x - 1)(2x - 9)$	$- 92x$

The complete problem is

$$20a^2 - 27a + 9 = (4x - 3)(5x - 3)$$

41. $10x^2 - 29x - 21$

Possible Factors	Middle Term When Multiplied
$(2x + 1)(5x - 21)$	$- 37x$
$(2x + 21)(5x - 1)$	$103x$
$(2x + 3)(5x - 7)$	x
$(2x + 7)(5x - 3)$	$29x$
$(x + 7)(10x - 3)$	$67x$
$(x + 3)(10x - 7)$	$23x$
$(x + 1)(10x - 21)$	$- 11x$
$(x + 21)(10x - 1)$	$209x$

The fourth factorization is correct but has the wrong signs so change signs. The complete problem is

$$10x^2 - 29x - 21 = (2x - 7)(5x + 3)$$

45. Prime

49. $64a^2 - 121b^2 = (8a)^2 - (11b)^2$
$$= (8a + 11b)(8a - 11b)$$

53. $y^2 + 20y + 100 = (y + 10)(y + 10)$
$$= (y + 10)^2$$

57. $16n^2 - 25n + 9 = (4n - 3)(4n - 3)$
$$= (4n - 3)^2$$

61. $9m^2 + 30mn + 25n^2 = (3m + 5n)(3m + 5n)$
$$= (3m + 5n)^2$$

65. $3m^3 - 18m^2 - 21m = 3m(m^2 - 6m - 7)$
$$= 3m(m + 1)(m - 7)$$

69. $8x^2 + 16x + 6 = 2(4x^2 + 8x + 3)$
$$= 2(2x + 1)(2x + 3)$$

73. $30x^2y - 55xy^2 + 15y^3 = 5y(6x^2 - 11xy + 3y^2)$
$$= 5y(3x - y)(2x - 3y)$$

77. $5x^2 - 45 = 5(x^2 - 9)$
$$= 5(x + 3)(x - 3)$$

81. $6a^3b + 33a^2b^2 + 15ab^3 = 3ab(2a^2 + 11ab + 5b^2)$
$$= 3ab(2a + b)(a + 5b)$$

85. $30a^4b + 35a^3b^2 - 15a^2b^3 = 5a^2b(6a^2 + 7ab - 3b^2)$
$$= 5a^2b(3a - b)(2a + 3b)$$

89. $3(2y + 5)(2y - 5) = 0$

$2y + 5 = 0$ or	$2y - 5 = 0$	Set each variable factor to 0
$2y = -5$	$2y = 5$	
$y = \dfrac{-5}{2}$	$y = \dfrac{5}{2}$	

The two solutions are $-\dfrac{5}{2}$ and $\dfrac{5}{2}$.

93.
$$a^2 - 49 = 0$$
$$(a + 7)(a - 7) = 0 \qquad \text{Factor completely}$$

$a + 7 = 0$ or	$a - 7 = 0$	Set each variable factor to 0
$a = -7$	$a = 7$	

The two solutions are -7 and 7.

97.
$$6y^2 = -13y - 6$$

$6y^2 + 13y + 6 = 0$		Add $13y + 6$ to each side
$(3y + 2)(2y + 3) = 0$		Factor completely

$3y + 2 = 0$ or	$2y + 3 = 0$	Set each variable factor to 0
$3y = -2$	$2y = -3$	
$y = -\dfrac{2}{3}$	$y = -\dfrac{3}{2}$	

101. Let x = the first even integer, then $x + 2$ = the second even integer.

$x(x + 2) = 120$	Their product is 120
$x^2 + 2x = 120$	Distributive property
$x^2 + 2x - 120 = 0$	Add -120 to each side
$(x - 10)(x + 12) = 0$	Factor completely

$x - 10 = 0$ or	$x + 12 = 0$	Set each variable factor to 0
$x = 10$	$x = -12$	First even integers
$x + 2 = 12$	$x + 2 = -14$	Second even integers

The solutions are 10,12 and -12,-10.

105. Let x = one number, then 2x - 1 = another number

$$x(2x - 1) = 66 \qquad \text{Their product is 66}$$
$$2x^2 - x = 66 \qquad \text{Distributive property}$$
$$2x^2 - x - 66 = 0 \qquad \text{Add -66 to each side}$$
$$(2x + 11)(x - 6) = 0 \qquad \text{Factor completely}$$

$$2x + 11 = 0 \quad \text{or} \quad x - 6 = 0 \qquad \text{Set each variable factor to 0}$$
$$2x = -11$$

$$x = -\frac{11}{2} \qquad\qquad x = 6 \qquad \text{one number}$$

$$2x - 1 = -12 \qquad 2x - 1 = 11 \qquad \text{another number}$$

The solutions are $-\frac{11}{2}$, -12 and 6, 11.

Chapter 5 Test

1. $5x - 10 = 5(x - 2)$

2. $18x^2y - 9xy - 36xy^2 = 9xy(2x - 1 - 4y)$

3. $\begin{aligned} x^2 + 2ax - 3bx - 6ab &= (x^2 + 2ax) + (-3bx - 6ab) \\ &= x(x + 2a) - 3b(x + 2a) \\ &= (x + 2a)(x - 3b) \end{aligned}$

4. $\begin{aligned} xy + 4x - 7y - 28 &= (xy + 4x) + (-7y - 28) \\ &= x(y + 4) - 7(y + 4) \\ &= (y + 4)(x - 7) \end{aligned}$

5. $x^2 - 5x + 6 = (x - 3)(x - 2)$

6. $x^2 - x - 6 = (x + 2)(x - 3)$

7. $a^2 - 16 = (a + 4)(a - 4)$

8. $x^2 + 25$ no factors - prime

9. $\begin{aligned} x^4 - 81 &= (x^2 + 9)(x^2 - 9) \\ &= (x^2 + 9)(x + 3)(x - 3) \end{aligned}$

10. $\begin{aligned} 27x^2 - 75y^2 &= 3(9x^2 - 25y^2) \\ &= 3(3x + 5y)(3x - 5y) \end{aligned}$

11. $\begin{aligned} x^3 + 5x^2 - 9x - 45 &= (x^3 + 5x^2) + (-9x - 45) \\ &= x^2(x + 5) - 9(x + 5) \\ &= (x + 5)(x^2 - 9) \\ &= (x + 5)(x + 3)(x - 3) \end{aligned}$

12. $\begin{aligned} x^2 - bx + 5x - 5b &= (x^2 - bx) + (5x - 5b) \\ &= x(x - b) + 5(x - b) \\ &= (x - b)(x + 5) \end{aligned}$

13. $\begin{aligned} 4a^2 + 22a + 10 &= 2(2a^2 + 11a + 5) \\ &= 2(2a + 1)(x + 5) \end{aligned}$

14. $\begin{aligned} 3m^2 - 3m - 18 &= 3(m^2 - m - 6) \\ &= 3(m + 2)(m - 3) \end{aligned}$

15. $6y^2 + 7y - 5 = (3y + 5)(2y - 1)$

16. $\begin{aligned} 12x^3 - 14x^2 - 10x &= 2x(6x^2 - 7x - 5) \\ &= 2x(3x + 5)(2y - 1) \end{aligned}$

17.
$$x^2 + 7x + 12 = 0$$
$$(x + 3)(x + 4) = 0$$

$x + 3 = 0$ or $x + 4 = 0$
      $x = -3$ $x = -4$

18.
$$x^2 + 4x + 4 = 0$$
$$(x - 2)(x - 2) = 0$$

$x - 2 = 0$ or $x - 2 = 0$
      $x = 2$ same

19.
$$x^2 - 36 = 0$$
$$(x + 6)(x - 6) = 0$$

$x + 6 = 0$ or $x - 6 = 0$
      $x = -6$ $x = 6$

20.
$$x^2 = x + 20$$
$$x^2 - x - 20 = 0$$
$$(x + 4)(x - 5) = 0$$

$x + 4 = 0$ or $x - 5 = 0$
      $x = -4$ $x = 5$

21.
$$x^2 - 11x = -30$$
$$x^2 - 11x + 30 = 0$$
$$(x - 5)(x - 6) = 0$$

$x - 5 = 0$ or $x - 6 = 0$
      $x = 5$ $x = 6$

22.
$$y^2 = 16y$$
$$y^3 - 16y = 0$$
$$y(y^2 - 16) = 0$$
$$y(y + 4)(y - 4) = 0$$

$y = 0$ or $y + 4 = 0$ or $y - 4 = 0$
                $y = -4$ $y = 4$

23.
$$2a^2 = a + 15$$
$$2a^2 - a - 15 = 0$$
$$(2a + 5)(a - 3) = 0$$

$2a + 5 = 0$ or $a - 3 = 0$
    $2a = -5$ $a = 3$

$$a = \frac{-5}{2}$$

24.
$$30x^3 - 20x^2 = 10x$$
$$30x^3 - 20x^2 - 10x = 0$$
$$10x(3x^2 - 2x - 1) = 0$$
$$10x(3x + 1)(x - 1) = 0$$

$10x = 0$ or $3x + 1 = 0$ or $x - 1 = 0$
$\quad x = 0 \qquad\qquad 3x = -1 \qquad\qquad x = 1$

$$x = \frac{-1}{3}$$

25. If we let x represent one of the numbers, then 20 - x must be the other number because their sum is 20. Since their product is 64, we can write

$$x(20 - x) = 64$$
$$20x - x^2 = 64$$
$$0 = x^2 - 20x + 64$$
$$0 = (x - 4)(x - 16)$$

$0 = x - 4$ or $0 = x - 16$
$\quad 4 = x \qquad\qquad 16 = x$
$16 = 20 - x \qquad 4 = 20 - x$

The solution is {4,16}.

26. Let x = the first of two consecutive odd integers, then x + 2 = the second of two consecutive odd integers.

$$x(x + 2) = (x + x + 2) + 7$$
$$x^2 + 2x = 2x + 9$$
$$x^2 - 9 = 0$$
$$(x + 3)(x - 3) = 0$$

$x + 3 = 0$ or $x - 3 = 0$
$\quad x = -3 \qquad\qquad x = 3$
$x + 2 = -1 \qquad x + 2 = 5$

We have two pairs of consecutive odd integers that are solutions. They are -3,-1 and 3,-5.

27. Let x = width and 3x + 5 = length.

Area = length · width

$$42 = (3x + 5)x$$
$$42 = 3x^2 + 5x$$
$$0 = 3x^2 + 5x - 42$$
$$0 = (x - 3)(3x + 14)$$

$0 = x - 3$ or $0 = 3x + 14$
$\quad 3 = x \qquad\qquad -14 = 3$

$$\frac{-14}{3} = x$$

The solution $x = \frac{-14}{3}$ cannot be used, since the length and width are always given in positive units.

Width = 3 feet
Length = 3x + 5 = 14 feet

Chapter 5 Test continued

28. Let x = the shortest side, the middle side = 2x + 2, the longest (hypotenuse) = 13 meters.

$$13^2 = x^2 + (2x + 2)^2 \qquad \text{Pythagorean Theorem}$$
$$169 = x^2 + 4x^2 + 8x + 4$$
$$0 = 5x^2 + 8x - 165$$
$$0 = (5x + 33)(x - 5)$$

$$0 = 5x + 33 \quad \text{or} \quad 0 = x - 5$$

$$\frac{-33}{5} = x \qquad\qquad 5 = x$$

The solution $\frac{-33}{5}$ cannot be used, since the length is always given in positive units. The lengths are 5 meters and 12 meters.

29.

$$C = 200 + 500x - 100x^2 \qquad \text{when } C = \$800$$
$$800 = 200 + 500x - 100x^2$$
$$100x^2 - 500x + 600 = 0$$
$$100(x^2 - 5x + 6) = 0$$
$$100(x - 2)(x - 3) = 0$$

$$x - 2 = 0 \quad \text{or} \quad x - 3 = 0$$
$$x = 2 \qquad\qquad x = 3$$

The solutions are 2 hundred items and 3 hundred items.

30. R = xp = (900 - 100p)p = \$1800

$$(900 - 100p)p = 1800$$
$$900p - 100p^2 = 1800$$
$$0 = 100p^2 - 900p + 1800$$
$$0 = 100(p^2 - 9p + 18)$$
$$0 = 100(p - 3)(p - 6)$$

$$0 = p - 3 \quad \text{or} \quad 0 = p - 6$$
$$3 = p \qquad\qquad 6 = p$$

The solutions are \$3 and \$6.

Chapter 6

Section 6.1

1. $\dfrac{5}{5x - 10} = \dfrac{5}{5(x - 2)}$

    $\qquad\quad = \dfrac{1}{x - 2} \qquad x = 2$

5. $\dfrac{a - 3}{a^2 - 9} = \dfrac{\cancel{a - 3}}{(a + 3)\cancel{(a - 3)}}$

    $\qquad\quad = \dfrac{1}{a + 3} \qquad\qquad a = -3, \ a = 3$

9. $\dfrac{2a}{10a^2} = \dfrac{2a}{10a^2}$

    $\qquad\quad = \dfrac{1}{5a} \qquad a = 0$

13. $\dfrac{35m^2}{7m^4} = \dfrac{\overset{5}{\cancel{35}}m^2}{\cancel{7}m^4}$

    $\qquad\quad = \dfrac{5}{m^2}$

17. $\dfrac{10a + 20}{5a + 10} = \dfrac{\overset{2}{\cancel{10}}(a + 2)}{\cancel{5}(a + 2)}$

    $\qquad\qquad = 2$

21. $\dfrac{x - 3}{x^2 - 6x + 9} = \dfrac{\cancel{x - 3}}{\cancel{(x - 3)}(x - 3)}$

    $\qquad\qquad = \dfrac{1}{x - 3}$

25. $\dfrac{3x + 5}{3x^2 + 24x + 45} = \dfrac{3(x + 5)}{3(x^2 + 8x + 15)}$

    $\qquad\qquad\quad = \dfrac{\cancel{3(x + 5)}}{\cancel{3(x + 5)}(x + 3)}$

    $\qquad\qquad\quad = \dfrac{1}{x + 3}$

29. $\dfrac{3x - 2}{9x^2 - 4} = \dfrac{\cancel{3x - 2}}{\cancel{(3x - 2)}(3x + 2)}$

    $\qquad\qquad = \dfrac{1}{3x + 2}$

Section 6.1 continued

33. $\dfrac{2m^3 - 2m^2 - 12m}{m^2 - 5m + 6} = \dfrac{m(m^2 - m - 6)}{m^2 - 5m + 6}$

$$= \dfrac{2m(m + 2)(m - 3)}{(m - 2)(m - 3)}$$

$$= \dfrac{2m(m + 2)}{m - 2}$$

37. $\dfrac{x^3 + 3x^2 - 4x}{x^3 - 16x} = \dfrac{x(x^2 + 3x - 4)}{x(x^2 - 16)}$

$$= \dfrac{x(x + 4)(x - 1)}{x(x + 4)(x - 4)}$$

$$= \dfrac{x - 1}{x - 4}$$

41. $\dfrac{4x^2 - 12x + 9}{4x^2 - 9} = \dfrac{(2x - 3)(2x - 3)}{(2x + 3)(2x - 3)}$

$$= \dfrac{2x - 3}{2x + 3}$$

45. $\dfrac{42x^3 - 20x^2 - 48x}{6x^2 - 5x - 4} = \dfrac{2x(21x^2 - 10x - 24)}{(2x + 1)(3x - 4)}$

$$= \dfrac{2x(7x + 6)(3x - 4)}{(2x + 1)(3x - 4)}$$

$$= \dfrac{2x(7x + 6)}{2x + 1}$$

49. $\dfrac{x^2 - 3x + ax - 3a}{x^2 - 3x + bx - 3b} = \dfrac{(x^2 - 3x) + (ax - 3a)}{(x^2 - 3x) + (bx - 3b)}$ Group terms

$$= \dfrac{x(x - 3) + a(x - 3)}{x(x - 3) + b(x - 3)} \quad \text{Factor}$$

$$= \dfrac{(x - 3)(x + a)}{(x - 3)(x + b)} \quad \text{Factor}$$

$$= \dfrac{x + a}{x - b} \quad \text{Simplify}$$

53. $\dfrac{x^3 + 2x^2 + 3x + 6}{x^2 - 3x - 10} = \dfrac{(x^3 + 2x^2) + (3x + 6)}{x^2 - 3x - 10}$ Group terms

$$= \dfrac{x^2(x + 2) + 3(x + 2)}{(x + 2)(x - 5)} \quad \text{Factor}$$

$$= \dfrac{(x + 2)(x^2 + 3)}{(x + 2)(x - 5)} \quad \text{Factor}$$

$$= \dfrac{x^2 + 3}{x - 5} \quad \text{Simplify}$$

157

Section 6.1 continued

57. $\dfrac{a - 7}{7 - a}$, Let $a = 10$, then

$$\frac{10 - 7}{7 - 10} = \frac{3}{-3} = -1$$

61. When $L - 630$, $d = 4$, $E = \dfrac{7L}{88d^2}$ becomes

$$E = \frac{7 \cdot 630}{88 \cdot 4^2}$$

$$E = \frac{4410}{88 \cdot 16}$$

$$E = \frac{4410}{1408}$$

$$E = \frac{2205}{704} \approx 3.1 \text{ lumens/square foot}$$

65. $\dfrac{3}{8} \cdot \dfrac{5}{2} = \dfrac{15}{16}$

69. $\dfrac{2}{3} \div \dfrac{1}{3} = \dfrac{2}{3} \cdot \dfrac{3}{1} = 2$

73. $75 = 3 \cdot 5 \cdot 5$
$\quad\quad = 3 \cdot 5^2$

Section 6.2

1. $\dfrac{x + y}{3} \cdot \dfrac{6}{x + y} = \dfrac{x + y}{\cancel{3}} \cdot \dfrac{\cancel{6}^{2}}{x + y}$
$$= 2$$

5. $\dfrac{9}{2a - 8} \div \dfrac{3}{a - 4} = \dfrac{\cancel{9}^{3}}{2(\cancel{a - 4})} \cdot \dfrac{\cancel{a - 4}}{\cancel{3}}$
$$= \dfrac{3}{2}$$

9. $\dfrac{a^2 + 5a}{7a} \cdot \dfrac{4a^2}{a^2 + 4a} = \dfrac{a(a + 5)}{7\cancel{a}} \cdot \dfrac{4a^{\cancel{2}}}{\cancel{a}(a + 4)}$
$$= \dfrac{4a(a + 5)}{7(a + 4)}$$

13. $\dfrac{2x - 8}{x^2 - 4} \cdot \dfrac{x^2 + 6x + 8}{x - 4} = \dfrac{2\cancel{(x - 4)}}{(x + 2)(x - 2)} \cdot \dfrac{(x + 2)(x + 4)}{\cancel{(x - 4)}}$
$$= \dfrac{2(x + 4)}{x - 2}$$

158

Section 6.2 continued

17. $\dfrac{a^2 + 10a + 25}{a + 5} \div \dfrac{a^2 - 25}{a - 5} = \dfrac{(a + 5)(a + 5)}{a + 5} \cdot \dfrac{a - 5}{(a + 5)(a - 5)}$

$\qquad\qquad\qquad = 1$

21. $\dfrac{2x^2 + 17x + 21}{x^2 + 2x - 35} \cdot \dfrac{x^2 - 25}{2x^2 - 7x - 15}$

$\qquad = \dfrac{(2x + 3)(x + 7)}{(x + 7)(x - 5)} \cdot \dfrac{(x + 5)(x - 5)}{(2x + 3)(x - 5)}$ Factor completely

$\qquad = \dfrac{(2x + 3)(x + 7)(x + 5)(x - 5)}{(x + 7)(x - 5)(2x + 3)(x - 5)}$ Divide out common factors

$\qquad = \dfrac{x + 5}{x - 5}$

25. $\dfrac{2a^2 + 7a + 3}{a^2 - 16} \div \dfrac{4a^2 + 8a + 3}{2a^2 - 5a - 12} = \dfrac{(2a + 1)(a + 3)}{(a - 4)(a + 4)} \cdot \dfrac{(2a + 3)(a - 4)}{(2a + 1)(2a + 3)}$

$\qquad\qquad\qquad\qquad = \dfrac{a + 3}{a - 4}$

29. $\dfrac{x^2 - 1}{6x^2 + 42x + 60} \cdot \dfrac{7x^2 + 17x + 6}{x + 1} \cdot \dfrac{6x + 30}{7x^2 - 11x - 6}$

$\qquad = \dfrac{(x + 1)(x - 1)}{6(x + 2)(x + 5)} \cdot \dfrac{(7x + 3)(x + 2)}{x + 1} \cdot \dfrac{6(x + 5)}{(7x + 3)(x - 2)}$ Factor completely

$\qquad = \dfrac{(x + 1)(x - 1)(7x + 3)(x + 2)6(x + 5)}{6(x + 2)(x + 5)(x + 1)(7x + 3)(x - 2)}$ Divide out common factors

$\qquad = \dfrac{x - 1}{x - 2}$

33. $(x^2 - 9)\left(\dfrac{2}{x + 3}\right) = \dfrac{x^2 - 9}{1} \cdot \dfrac{2}{x + 3}$

$\qquad\qquad\qquad = \dfrac{(x + 3)(x - 3)2}{x + 3}$

$\qquad\qquad\qquad = 2(x - 3)$

37. $(x^2 - x - 6)\left(\dfrac{x + 1}{x - 3}\right) = \dfrac{x^2 - x - 6}{1} \cdot \dfrac{x + 1}{x - 3}$

$\qquad\qquad\qquad = \dfrac{(x - 3)(x + 2)(x + 1)}{x - 3}$

$\qquad\qquad\qquad = (x + 2)(x + 1)$

41. $\dfrac{x^2 - 9}{x^2 - 3x} \cdot \dfrac{2x + 10}{xy + 5y + 3y + 15}$

$$= \dfrac{(x + 3)(x - 3)}{x(x - 3)} \cdot \dfrac{2(x + 5)}{(xy + 5x) + (3y + 15)}$$

$$= \dfrac{(x + 3)(x - 3)}{x(x - 3)} \cdot \dfrac{2(x + 5)}{x(y + 5) + 3(y + 5)}$$

$$= \dfrac{(x + 3)(x - 3)}{x(x - 3)} \cdot \dfrac{2(x + 5)}{(x + 3)(y + 5)}$$

$$= \dfrac{2(x + 5)}{x(y + 5)}$$

45. $\dfrac{x^3 - 3x^2 + 4x - 12}{x^4 - 16} \cdot \dfrac{3x^2 + 5x - 2}{3x^2 - 10x + 3}$

$$= \dfrac{x^2(x - 3) + 4(x - 3)}{(x^2 + 4)(x^2 - 4)} \cdot \dfrac{(x + 2)(3x - 1)}{(3x - 1)(x - 3)} \qquad \text{Factor}$$

$$= \dfrac{(x - 3)(x^2 + 4)}{(x^2 + 4)(x + 2)(x - 2)} \cdot \dfrac{(x + 2)(3x - 1)}{(3x - 1)(x - 3)} \qquad \text{Factor}$$

$$= \dfrac{(x - 3)(x^2 + 4)(x + 2)(3x - 1)}{(x^2 + 4)(x + 2)(x - 2)(3x - 1)(x - 3)} \qquad \begin{array}{l}\text{Divide out} \\ \text{common factors}\end{array}$$

$$= \dfrac{1}{x - 2}$$

49. $\left(1 - \dfrac{1}{2}\right)\left(1 - \dfrac{1}{3}\right)\left(1 - \dfrac{1}{4}\right)\ldots\left(1 - \dfrac{1}{99}\right)\left(1 - \dfrac{1}{100}\right)$

$$= \left(\dfrac{1}{2}\right)\left(\dfrac{2}{3}\right)\left(\dfrac{3}{4}\right)\ldots\left(\dfrac{98}{99}\right)\left(\dfrac{99}{100}\right) \qquad \text{Subtract}$$

$$= \left(\dfrac{1}{2}\right)\left(\dfrac{2}{3}\right)\left(\dfrac{3}{4}\right)\ldots\left(\dfrac{98}{99}\right)\left(\dfrac{99}{100}\right) \qquad \begin{array}{l}\text{Divide out} \\ \text{common factors}\end{array}$$

$$= \dfrac{1}{100}$$

53. $\dfrac{1}{10} + \dfrac{3}{14} = \dfrac{1 \cdot 7}{10 \cdot 7} + \dfrac{3 \cdot 5}{14 \cdot 5} \qquad \text{LCD} = 2 \cdot 5 \cdot 7$

$$= \dfrac{7}{70} + \dfrac{15}{70}$$

$$= \dfrac{22}{70}$$

$$= \dfrac{11}{35}$$

Section 6.2 continued

57. $\dfrac{10x^4}{2x^2} + \dfrac{12x^6}{3x^4}$

$\qquad = 5x^2 + 4x^2$

$\qquad = 9x^2$

Section 6.3

1. $\dfrac{3}{x} + \dfrac{4}{x} = \dfrac{7}{x}$

5. $\dfrac{1}{x+1} + \dfrac{x}{x+1} = \dfrac{x+1}{x+1} = 1$

9. $\dfrac{x}{x+2} + \dfrac{4x+4}{x+2} = \dfrac{x^2+4x+4}{x+2}$

$\qquad\qquad = \dfrac{(x+2)\cancel{(x+2)}}{\cancel{x+2}}$

$\qquad = x + 2$

13. $\dfrac{x+2}{x+6} - \dfrac{x-4}{x+6} = \dfrac{x+2-x-4}{x+6} = \dfrac{6}{x+6}$

17. $\dfrac{1}{2} + \dfrac{a}{3} = \dfrac{3}{6} + \dfrac{2a}{6} = \dfrac{3+2a}{6}$

21. $\dfrac{x+1}{x-2} - \dfrac{4x+7}{5x-10} = \dfrac{x+1}{x-2} - \dfrac{4x+7}{5(x-2)}$

$\qquad\qquad = \dfrac{5}{5}\left(\dfrac{x+1}{x-2}\right) - \dfrac{4x+7}{5(x-2)}$

$\qquad\qquad = \dfrac{5x+5}{5(x-2)} - \dfrac{4x+7}{5(x-2)}$

$\qquad\qquad = \dfrac{\cancel{x-2}}{5\cancel{(x-2)}} = \dfrac{1}{5}$

25. $\dfrac{6}{x(x-2)} + \dfrac{3}{x}$

$\qquad = \dfrac{6}{x(x-2)} + \dfrac{3(x-2)}{x(x-2)}$

$\qquad = \dfrac{6+3x-6}{x(x-2)}$

$\qquad = \dfrac{3x}{x(x-2)}$

$\qquad = \dfrac{3}{x-2}$

29. $\dfrac{2}{x + 5} - \dfrac{10}{x^2 - 25}$

$$= \dfrac{2}{x + 5} - \dfrac{10}{(x + 5)(x - 5)}$$

$$= \dfrac{2(x - 5)}{(x + 5)(x - 5)} - \dfrac{10}{(x + 5)(x - 5)}$$

$$= \dfrac{2x - 10 - 10}{(x + 5)(x - 5)}$$

$$= \dfrac{2x - 20}{(x + 5)(x - 5)}$$

33. $\dfrac{a - 4}{a - 3} + \dfrac{5}{a^2 - a - 6}$

$$= \dfrac{a - 4}{a - 3} + \dfrac{5}{(a + 2)(a - 3)}$$

$$= \dfrac{a + 2}{a + 2}\left(\dfrac{a - 4}{a - 3}\right) + \dfrac{5}{(a + 2)(a - 3)}$$

$$= \dfrac{a^2 - 2a - 8}{(a + 2)(a - 3)} + \dfrac{5}{(a + 2)(a - 3)}$$

$$= \dfrac{a^2 - 2a - 3}{(a + 2)(a - 3)}$$

$$= \dfrac{(a + 1)(a - 3)}{(a + 2)(a - 3)}$$

$$= \dfrac{a + 1}{a + 2}$$

37. $\dfrac{4y}{y^2 + 6y + 5} - \dfrac{3y}{y^2 + 5y + 4}$

$$= \dfrac{4y}{(y + 5)(y + 1)} - \dfrac{3y}{(y + 4)(y + 1)}$$

$$= \dfrac{4y(y + 4)}{(y + 5)(y + 1)(y + 4)} - \dfrac{3y(y + 5)}{(y + 5)(y + 1)(y + 4)}$$

$$= \dfrac{4y^2 + 16y - 3y^2 - 15y}{(y + 5)(y + 1)(y + 4)}$$

$$= \dfrac{y^2 + y}{(y + 5)(y + 1)(y + 4)}$$

$$= \dfrac{y(y + 1)}{(y + 5)(y + 1)(y + 4)}$$

$$= \dfrac{y}{(y + 5)(y + 4)}$$

41. $\dfrac{1}{x} + \dfrac{x}{3x + 9} - \dfrac{3}{x^2 + 3x}$

$$= \dfrac{1}{x} + \dfrac{x}{3(x + 3)} - \dfrac{3}{x(x + 3)}$$

$$= \dfrac{3(x + 3)}{3(x + 3)} \left(\dfrac{1}{x} \right) + \dfrac{x}{x} \left(\dfrac{x}{3(x + 3)} \right) - \dfrac{3}{3} \left(\dfrac{3}{x(x + 3)} \right)$$

$$= \dfrac{3x + 9}{3x(x + 3)} + \dfrac{x^2}{3x(x + 3)} - \dfrac{9}{3x(x + 3)}$$

$$= \dfrac{x^2 + 3x}{3x(x + 3)}$$

$$= \dfrac{x(x + 3)}{3x(x + 3)}$$

$$= \dfrac{1}{3}$$

45. $1 - \dfrac{1}{x + 1} = \dfrac{x + 1}{x + 1} (1) - \dfrac{1}{x + 1}$

$$= \dfrac{x + 1}{x + 1} - \dfrac{1}{x + 1}$$

$$= \dfrac{x + 1 - 1}{x + 1}$$

$$= \dfrac{x}{x + 1}$$

49. $1 - \dfrac{1}{x + 3} = \dfrac{x + 3}{x + 3} (1) - \dfrac{1}{x + 3}$

$$= \dfrac{x + 3}{x + 3} - \dfrac{1}{x + 3}$$

$$= \dfrac{x + 3 - 1}{x + 3}$$

$$= \dfrac{x + 2}{x + 3}$$

53. $\dfrac{1}{x} + \dfrac{1}{2x}$

$$= \dfrac{2 \cdot 1}{2x} + \dfrac{1}{2x}$$

$$= \dfrac{3}{2x}$$

Section 6.3 continued

57. x - 3(x + 3) = x - 3
      x - 3x - 9 = x - 3
         -2x - 9 = x - 3   Add 2x to both sides
            -9 = 3x - 3
            -6 = 3x   Add 3 to both sides
            -2 = x   Multiply both sides by $\frac{1}{3}$

61. $x^2 + 5x + 6 = 0$
     (x + 2)(x + 3) = 0   Factor the left side

  x + 2 = 0   or   x + 3 = 0   Set factor to 0
    x = -2   x = -3   Solutions

65. $x^2 - 5x = 0$
     x(x - 5) = 0   Factor the left side

  x = 0   or   x - 5 = 0   Set factors to 0
  x = 0   x = 5   Solutions

Section 6.4

1. $\frac{x}{3} + \frac{1}{2} = -\frac{1}{2}$

$6(\frac{x}{3}) + 6(\frac{1}{2}) = 6(-\frac{1}{2})$ Multiply both sides by 6

    2x + 3 = -3
       2x = -6
        x = -3

The solution set is {-3}.

5. $\frac{3}{x} + 1 = \frac{2}{x}$

$x(\frac{3}{x}) + x(1) = x(\frac{2}{x})$ Multiply both sides by x

    3 + x = 2
       x = -1

The solution set is {-1}.

9. $\frac{3}{x} + 2 = \frac{1}{2}$

$2x(\frac{3}{x}) + 2x(2) = 2x(\frac{1}{2})$

    6 + 4x = x
        6 = -3x
      -2 = x

The solution set is {-2}.

13.
$$1 - \frac{8}{x} = \frac{-15}{x^2}$$

$$x^2(1) - x^2\left(\frac{8}{x}\right) = x^2\left(\frac{-15}{x^2}\right)$$

$$x^2 - 8x = -15$$
$$x^2 - 8x + 15 = 0$$
$$(x - 3)(x - 5) = 0$$

$$x - 3 = 0 \quad \text{or} \quad x - 5 = 0$$
$$x = 3 \qquad\qquad x = 5$$

The solutions are 3 and 5.

17.
$$\frac{x - 3}{2} + \frac{2x}{3} = \frac{5}{6}$$

$$6\left(\frac{x - 3}{2}\right) + 6\left(\frac{2x}{2}\right) = 6\left(\frac{5}{6}\right)$$

$$3(x - 3) + 2(2x) = 5$$
$$3x - 9 + 4x = 5$$
$$7x - 9 = 5$$
$$7x = 14$$

$$x = \frac{14}{7}$$

$$x = 2$$

The solution set is $\{2\}$.

21.
$$\frac{6}{x + 2} = \frac{3}{5}$$

$$5(x + 2)\left(\frac{6}{x + 2}\right) = 5(x + 2)\left(\frac{3}{5}\right)$$

$$30 = 3(x + 2)$$
$$30 = 3x + 6$$
$$24 = 3x$$
$$8 = x$$

The solution set is $\{8\}$.

25.
$$\frac{x}{x - 2} + \frac{2}{3} = \frac{2}{x - 2}$$

$$3(x - 2)\left(\frac{x}{x - 2}\right) + 3(x - 2)\left(\frac{2}{3}\right) = 3(x - 2)\left(\frac{2}{x - 2}\right)$$

$$3x + 2(x - 2) = 6$$
$$3x + 2x - 4 = 6$$
$$5x = 10$$
$$x = 2$$

Possible solution 3, which does not check, $\emptyset$.

165

29.

$$\frac{5}{x + 2} + \frac{1}{x + 3} = \frac{-1}{x^2 + 5x + 6}$$

$$(x + 2)(x + 3)\frac{5}{x + 2} + (x + 2)(x + 3)\frac{1}{x + 3} = (x + 2)(x + 3)\frac{-1}{(x + 2)(x + 3)}$$

$$5(x + 3 + x + 2) = -1$$
$$5x + 15 + x + 2 = -1$$
$$6x + 17 = -1$$
$$6x = -18$$
$$x = -3$$

Possible solution -3, which does not check; $\emptyset$.

33.

$$\frac{a}{2} + \frac{3}{a - 3} = \frac{a}{a - 3}$$

$$2(a - 3)\frac{a}{2} + 2(a - 3)(\frac{3}{a - 3}) = 2(a - 3)\frac{a}{a - 3}$$

$$a(a - 3) + 6 = 2a$$
$$a^2 - 3a + 6 = 2a$$
$$a^2 - 5a + 6 = 1$$
$$(a - 2)(a - 3) = 0$$

$$a - 2 = 0 \quad \text{or} \quad a - 3 = 0$$
$$a = 2 \qquad\qquad a = 3$$

Possible solutions 2 and 3, but only 2 checks; 2.

37.

$$\frac{2}{a^2 - 9} = \frac{3}{a^2 + a - 12}$$

$$\frac{2}{(a + 3)(a - 3)} = \frac{3}{(a + 4)(a - 3)}$$

$$(a + 3)(a - 3)(a + 4) \cdot \frac{2}{(a + 3)(a - 3)} = (a + 3)(a - 3)(a + 4) \cdot \frac{3}{(a + 4)(a - 3)}$$

$$2(a + 4) = 3(a + 3)$$
$$2a + 8 = 3a + 9$$
$$8 = a + 9$$
$$-1 = a$$

The solution set is {-1}.

41.

$$\frac{2x}{x + 2} = \frac{x}{x + 3} - \frac{3}{x^2 + 5x + 6}$$

$$(x + 2)(x + 3)\frac{2x}{x + 2} = (x + 2)(x + 3)\frac{x}{x + 3} - (x + 2)(x + 3)\frac{3}{(x + 2)(x + 3)}$$

$$2x(x + 3) = x(x + 2) - 3$$
$$2x^2 + 6x = x^2 + 2x - 3$$
$$x^2 + 4x + 3 = 0$$
$$(x + 3)(x + 1) = 0$$

$$x + 3 = 0 \quad \text{or} \quad x + 1 = 0$$

Possible solutions -3 and -1, but only -1 checks; -1.

Section 6.4 continued

45.   Step 1:   Let x = the width

      Step 2:   Let $2x + 5$ = length

$$A = 2L + 2W$$
$$34 = 2(2x + 5) + 2x$$

      Step 3:

$$34 = 4x + 10 + 2x$$
$$34 = 6x + 10$$
$$24 = 6x$$
$$4 = x$$

Width is 4 inches, length is 13 inches.

49.

$10^2 = x^2 + (x + 2)^2$	Pythagorean theorem
$100 = x^2 + x^2 + 4x + 4$	Expand $(x + 2)^2$
$100 = 2x^2 + 4x + 4$	Simplify the right side
$0 = 2x^2 + 4x - 96$	Add -100 to both sides
$0 = 2(x^2 + 2x - 48)$	Begin factoring
$0 = 2(x + 8)(x - 6)$	Factor completely
$0 = x + 8$ or $0 = x - 6$	Set variable factors to 0
$-8 = x$ $6 = x$	

The two legs are 6 inches and 8 inches.

Section 6.5

1.   Let x = a number, then $3x$ is the other number.   Their reciprocals are $\frac{1}{x}$ and $\frac{1}{3x}$.

$$\frac{1}{x} + \frac{1}{3x} = \frac{16}{3}$$

$$3x\left(\frac{1}{x}\right) + 3x\left(\frac{1}{3x}\right) = 3x\left(\frac{16}{3}\right)$$

$$3 + 1 = 16x$$
$$4 = 16x$$

$$\frac{4}{16} = x$$

$$\frac{1}{4} = x$$

$$\frac{3}{4} = 3x$$

The numbers are $\frac{1}{4}, \frac{3}{4}$.

Section 6.5 continued

5. Let x = a certain number.

$$\frac{7 + x}{9 + x} = \frac{5}{7}$$

$$7(9 + x)\frac{7 + x}{9 + x} = 7(9 + x)\frac{5}{7}$$

$$7(7 + x) = 5(9 + x)$$
$$49 + 7x = 45 + 5x$$
$$49 + 2x = 45$$
$$2x = -4$$
$$x = -2$$

The number is -2.

9.

	d	r	t
upstream	26	x + 3	
downstream	38	x - 3	

x = the speed of the boat

To fill the two time blocks, remember d = rt. We use the formula $t = \frac{d}{r}$.

	d	r	t
upstream	26	x - 3	$\frac{26}{x - 3}$
downstream	38	x + 3	$\frac{38}{x + 3}$

The time upstream and the time downstream are equal.

$$\frac{26}{x - 3} = \frac{38}{x + 3}$$

$$(x + 3)(x - 3)\frac{26}{x - 3} = (x + 3)(x - 3)\frac{38}{x + 3} \qquad \text{LCD} = (x + 3)(x - 3)$$

$$(x + 3)26 = (x - 3)38$$
$$26x + 78 = 38x - 114$$
$$78 = 12x - 114$$
$$192 = 12x$$
$$16 = x$$

The speed of the boat is 16 mph.

13.

	d	r	t
faster plane	285	x + 20	
slower plane	225	x	

x = the slower plane

To fill the two time blocks, remember d = rt. We use the formula $t = \dfrac{d}{r}$.

	d	r	t
faster plane	285	x + 20	$\dfrac{285}{x + 20}$
slower plane	255	x	$\dfrac{255}{x}$

$$\frac{285}{x + 20} = \frac{255}{x}$$

$$(x + 20)x\frac{285}{x + 20} = (x + 20)x\frac{255}{x} \qquad LCD = (x + 20)x$$

$$285x = 255(x + 20)$$
$$285x = 255x + 5100$$
$$30x = 5100$$
$$x = 170$$

The slower plane has a speed of 170 mph and the faster plane has a speed of 190 mph.

17. Let x = amount of time to fill the tank with both pipes open.

$$\frac{1}{10} \quad + \quad \frac{1}{12} \quad = \quad \frac{1}{x}$$

Amount of time   Amount of time   Amount of time
for cold water   for hot water   for both waters

$$60x\left(\frac{1}{10}\right) + 60x\left(\frac{1}{12}\right) = 60x\left(\frac{1}{x}\right) \qquad LCD = 60x$$

$$6x + 5x = 60$$
$$11x = 60$$
$$x = \frac{60}{11}$$

It will take $\dfrac{60}{11}$ minutes to fill the tub if both faucets are open.

21. $y = \dfrac{-4}{x}$

x	y
-4	1
-2	2
-1	4
1	-4
2	-2
4	-1

See the graph on page A-28 in the textbook.

25. $y = \dfrac{3}{x}$ $x + y = 4$

x	y
1	3
3	1
-1	-3
-3	-1

x	y
0	4
1	3
3	1
4	0

The points of intersection are (1,3) and (3,1). See the graph on page A-28 in the textbook.

29.
$$
\begin{array}{l}
4x - 5y = 1 \xrightarrow{\text{no change}} \quad 4x - 5y = 1 \\
x - 2y = -2 \xrightarrow{\text{multiply by -4}} -4x + 8y = 8 \\
\hline
\qquad\qquad\qquad\qquad\qquad 0 + 3y = 9 \\
\qquad\qquad\qquad\qquad\qquad\quad 3y = 9 \\
\qquad\qquad\qquad\qquad\qquad\quad\; y = 9
\end{array}
$$

Substituting $y = 3$ into either of the two original equations, we get $x = 4$. The solution to the system is (4,3). It satisfies both equations.

33.
$$
\begin{array}{r}
2x - 3y = 4 \\
x = 2y + 1 \\
2(2y + 1) - 3y = 4 \\
4y + 2 - 3y = 4 \\
y + 2 = 4 \\
y = 2
\end{array}
$$

Substituting $y = 2$ into either of the two original equations, we get $x = 5$. The solution to the system is (5,2). It satisfies both equations.

1. Method 1 $\quad \dfrac{\frac{3}{4}}{\frac{1}{8}} = \dfrac{8 \cdot \frac{3}{4}}{8 \cdot \frac{1}{8}} \qquad$ LCD = 8

$$= \frac{6}{1}$$

$$= 6$$

Method 2 $\quad \dfrac{\frac{3}{4}}{\frac{1}{8}} = \dfrac{3}{4} \cdot \dfrac{8}{1}$

$$= \frac{6}{1}$$

$$= 6$$

5. Method 1 $\quad \dfrac{\frac{x^2}{y}}{\frac{x}{y^3}} = \dfrac{y^3 \cdot \frac{x^2}{y}}{y^3 \cdot \frac{x}{y^3}} \qquad$ LCD = y^3

$$= \frac{x^2 y^2}{x}$$

$$= xy^2$$

Method 2 $\quad \dfrac{\frac{x^2}{y}}{\frac{x}{y^3}} = \dfrac{x^2}{y} \cdot \dfrac{y^3}{x}$

$$= xy^2$$

9. $\quad \dfrac{y + \frac{1}{x}}{x + \frac{1}{y}} = \dfrac{xy\left(y + \frac{1}{x}\right)}{xy\left(x + \frac{1}{y}\right)} \qquad$ LCD = xy

$$= \frac{xy \cdot y + xy \cdot \frac{1}{x}}{xy \cdot x + xy \cdot \frac{1}{y}}$$

$$= \frac{xy^2 + y}{x^2 y + x}$$

$$= \frac{y(xy + 1)}{x(xy + 1)}$$

$$= \frac{y}{x}$$

Section 6.6 continued

13. Method 1 $\quad \dfrac{\dfrac{x+1}{x^2-9}}{\dfrac{2}{x+3}} = \dfrac{(x+3)(x-3)\left(\dfrac{x+1}{x^2-9}\right)}{(x+3)(x-3)\left(\dfrac{2}{x+3}\right)} \qquad$ LCD $= (x+3)(x-3)$

$$= \dfrac{x+1}{2(x-3)}$$

Method 2 $\quad \dfrac{\dfrac{x+1}{x^2-9}}{\dfrac{2}{x+3}} = \dfrac{x+1}{x^2-9} \cdot \dfrac{x+3}{2}$

$$= \dfrac{x+1}{(x+3)(x-3)} \cdot \dfrac{x+3}{2}$$

$$= \dfrac{x+1}{2(x-3)}$$

17. $\quad \dfrac{1-\dfrac{9}{y^2}}{1-\dfrac{1}{y}-\dfrac{6}{y^2}} = \dfrac{y^2\left(1-\dfrac{9}{y^2}\right)}{y^2\left(1-\dfrac{1}{y}-\dfrac{6}{y^2}\right)} \qquad$ LCD $= y^2$

$$= \dfrac{y^2-9}{y^2-y-6}$$

$$= \dfrac{(y+3)(y-3)}{(y-3)(y+2)}$$

$$= \dfrac{y+3}{y+2}$$

21. $\quad \dfrac{1-\dfrac{1}{a^2}}{1-\dfrac{1}{a}} = \dfrac{a^2\left(a-\dfrac{1}{}\right)}{a^2\left(1-\dfrac{1}{a}\right)} \qquad$ LCD $= a^2$

$$= \dfrac{a^2-1}{a^2-a}$$

$$= \dfrac{(a+1)(a-1)}{a(a-1)}$$

$$= \dfrac{a+1}{a}$$

25. $\quad \dfrac{\dfrac{1}{a+1}+2}{\dfrac{1}{a+1}+3} = \dfrac{a+1\left(\dfrac{1}{}+2\right)}{a+1\left(\dfrac{1}{a+1}+3\right)} \qquad$ LCD $= a+1$

$$= \dfrac{1+2(a+1)}{1+3(a+1)}$$

$$= \dfrac{1+2a+2}{1+3a+3}$$

$$= \dfrac{2a+3}{3a+4}$$

Section 6.6 continued

29. $(1 + \frac{1}{x + 3})(1 + \frac{1}{x + 2})(1 + \frac{1}{x + 1})$

$= (\frac{x + 3 + 1}{x + 3})(\frac{x + 2 + 1}{x + 2})(\frac{x + 1 + 1}{x + 1})$ Common demonimators

$= (\frac{x + 4}{x + 3})(\frac{x + 3}{x + 2})(\frac{x + 2}{x + 1})$ Add

$= (\frac{x + 4}{x + 3})(\frac{\cancel{x + 3}}{\cancel{x + 2}})(\frac{\cancel{x + 2}}{x + 1})$ Divide out common factors

$= \frac{x + 4}{x + 1}$

33. $-3x \leq 21$

$x \geq -7$ Remember to reverse the inequality symbol when multiplying by a negative number.

37. $4 - 2(x + 1) \geq -2$

$4 - 2x - 2 \geq -2$

$-2x + 2 \geq -2$

$-2x \geq -4$

$x \leq 2$ Remember to reverse the inequality symbol when multiplying by a negative number.

Section 6.7

1. 8 to 6 $\frac{8}{6} = \frac{4}{3}$

5. 32 to 4 $\frac{32}{4} = \frac{8}{1}$

9. $\frac{2}{3}$ to $\frac{3}{4}$ $\frac{\frac{2}{3}}{\frac{3}{4}} = \frac{2}{3} \cdot \frac{4}{3}$

$= \frac{8}{9}$

13. White bread / Whole wheat bread $\frac{60}{55} = \frac{12}{11}$

17. $\frac{20 \text{ minutes}}{2 \text{ hours}} = \frac{20 \text{ minutes}}{120 \text{ minutes}}$

$= \frac{1}{6}$

21. $\dfrac{2}{5} = \dfrac{4}{x}$ Extremes are 2 and x; means are 5 and 4

$2x = 20$ Product of extremes = product of means
$x = 10$ Divide both sides by 2

25. $\dfrac{a}{3} = \dfrac{5}{12}$ Extremes are a and 12; means are 3 and 5

$12a = 15$ Product of extremes = product of means

$a = \dfrac{15}{12}$ Divide both sides by 12

$a = \dfrac{5}{4}$ Reduce to lowest terms

29. $\dfrac{x + 1}{3} = \dfrac{4}{x}$ Extremes are x + 1 and x; means are 3 and 4

$x(x + 1) = 12$ Product of extremes = product of means
$x^2 + x = 12$
$x^2 + x - 12 = 0$ Standard form for a quadratic equation
$(x + 4)(x - 3) = 0$ Factor

$x + 4 = 0$ or $x - 3 = 0$ Set factors equal to 0
$x = -4$ $x = 3$

The solutions are -4 and 3.

33. $\dfrac{4}{x + 2} = \dfrac{a}{2}$ Extremes are 4 and 2; means are a + 2 and 2

$8 = a(a + 2)$ Product of extremes = product of means

$8 = a^2 + 2a$
$0 = a^2 + 2a - 8$ Standard form for a quadratic equation

$0 = (a + 4)(a - 2)$ Factor

$a + 4 = 0$ or $a - 2 = 0$ Set factors equal to 0
$a = -4$ $a = 2$

The solutions are -4 and 2.

37. $\dfrac{\text{hits}}{\text{games}} \dfrac{6}{18} = \dfrac{x}{45}$ Extremes are 6 and 45; means are 18 and x

$6(45) = 18x$ Product of extremes = product of means
$270 = 18x$

$\dfrac{270}{18} = x$ Divide both sides by 18
$15 = x$ Reduce to lowest terms

15 hits in the first 45 games

Section 6.7 continued

41. $\dfrac{\text{grams of ice cream}}{\text{grams of fat}} \quad \dfrac{100}{13} = \dfrac{350}{x}$ Extremes are 100 and x; means are 13 and 350

$$100x = 13(350)$$ Product of extremes = product or means

$$100x = 4550$$

$$x = 45.5$$ Grams of fat

45. $\dfrac{\text{miles}}{\text{hours}} \quad \dfrac{245}{5} = \dfrac{x}{7}$ Extremes are 245 and 7; means are 5 and x

$$245(7) = 5x$$ Product of extremes = product of means

$$1715 = 5x$$

$$343 = x$$

343 miles in 7 hours

49. $\dfrac{x^2 - 25}{x + 4} \cdot \dfrac{2x + 8}{x^2 - 9x + 20}$

$$= \dfrac{(x + 5)(x - 5)}{x + 4} \cdot \dfrac{2(x + 4)}{(x - 4)(x - 5)}$$

$$= \dfrac{2(x + 5)}{x - 4}$$

Chapter 6 Review

1. $\dfrac{7}{14x - 28} = \dfrac{7}{14(x - 3)}$

   $\qquad\qquad = \dfrac{1}{2(x - 2)} \qquad x \neq 2$

5. $\dfrac{8x - 4}{4x + 12} = \dfrac{4(2x - 1)}{4(x + 3)}$

   $\qquad\qquad = \dfrac{2x - 1}{x + 3} \qquad x \neq -3$

9. $\dfrac{3x^3 + 16x^2 - 12x}{2x^3 + 9x^2 - 18x} = \dfrac{x(3x^2 + 16x - 12)}{x(2x^2 + 9x - 18)}$ $\qquad$ Factor

   $\qquad\qquad = \dfrac{x(x + 6)(3x - 2)}{x(x + 6)(2x - 3)}$ $\qquad$ Factor

   $\qquad\qquad = \dfrac{\cancel{x}(\cancel{x + 6})(3x - 2)}{\cancel{x}(\cancel{x + 6})(2x - 3)}$

   $\qquad\qquad = \dfrac{3x - 2}{2x - 3}$

13. $\dfrac{x^2 + 5x - 14}{x + 7} = \dfrac{(x + 7)(x - 2)}{x + 7}$

    $\qquad\qquad = \dfrac{(\cancel{x + 7})(x - 2)}{\cancel{x + 7}}$

    $\qquad\qquad = x - 2$

17. $\dfrac{xy + bx + ay + ab}{xy + 5x + ay + 5a} = \dfrac{(xy + bx) + (ay + ab)}{(xy + 5x) + (ay + 5a)}$

    $\qquad\qquad = \dfrac{x(y + b) + (a(y + b)}{x(y + 5) + a(y + 5)}$

    $\qquad\qquad = \dfrac{(y + b)(x + a)}{(y + 5)(x + a)}$

    $\qquad\qquad = \dfrac{(y + b)(\cancel{x + a})}{(y + 5)(\cancel{x + a})}$

    $\qquad\qquad = \dfrac{y + b}{y + 5}$

Chapter 6 Review continued

21. $\dfrac{x^2 + 8x + 16}{x^2 + x - 12} \div \dfrac{x^2 - 16}{x^2 - x - 6}$

$= \dfrac{x^2 + 8x + 16}{x^2 + x - 12} \cdot \dfrac{x^2 - x - 6}{x^2 - 16}$ Division is multiplication by the reciprocal

$= \dfrac{(x + 4)(x + 4)}{(x + 4)(x - 3)} \cdot \dfrac{(x - 3)(x + 2)}{(x + 4)(x - 4)}$

$= \dfrac{(x + 4)(x + 4)(x - 3)(x + 2)}{(x + 4)(x - 3)(x + 4)(x - 4)}$ Factor and multiply

$= \dfrac{x + 2}{x - 4}$ Divide out common factors

25. $\dfrac{3x^2 - 2x - 1}{x^2 + 6x + 8} \div \dfrac{3x^2 + 13x + 4}{x^2 + 8x + 16}$

$= \dfrac{3x^2 - 2x - 1}{x^2 + 6x + 8} \cdot \dfrac{x^2 + 8x + 16}{3x^2 + 13x + 4}$ Division is multiplication by the reciprocal

$= \dfrac{(3x + 1)(x - 1)}{(x + 2)(x + 4)} \cdot \dfrac{(x + 4)(x + 4)}{(3x + 1)(x + 4)}$

$= \dfrac{(3x + 1)(x - 1)(x + 4)(x + 4)}{(x + 2)(x + 4)(3x + 1)(x + 4)}$ Factor and multiply

$= \dfrac{x - 1}{x + 2}$

29. $\dfrac{x^2}{x - 9} - \dfrac{18x - 81}{x - 9} = \dfrac{x^2 - 18x + 81}{x - 9}$ Subtract

$= \dfrac{(x - 9)(x - 9)}{x - 9}$

$= x - 9$

33. $\dfrac{x}{x + 9} + \dfrac{5}{x} = \dfrac{x \cdot x}{(x + 9)x} + \dfrac{5(x + 9)}{x(x + 9)}$ LCD $= x(x + 9)$

$= \dfrac{x^2 + 5x + 45}{x(x + 9)}$

37. $\dfrac{3}{x^2 - 36} - \dfrac{2}{x^2 - 4x - 12}$

$= \dfrac{3}{(x + 6)(x - 6)} - \dfrac{2}{(x + 2)(x - 6)}$ $\qquad$ LCD $= (x + 6)(x - 6)(x + 2$

$= \dfrac{3(x = 2)}{(x + 6)(x - 6)(x + 2)} - \dfrac{2(x + 6)}{(x + 2)(x - 6)(x + 6)}$

$= \dfrac{3x + 6 - 2x - 12}{(x + 6)(x - 6)(x + 2)}$

$= \dfrac{x - 6}{(x + 6)(x - 6)(x + 2)}$

$= \dfrac{1}{(x + 6)(x + 2)}$

41. $\qquad\qquad \dfrac{3}{x} + \dfrac{1}{2} = \dfrac{5}{x}$ $\qquad$ LCD $= 2x$

$2x \quad \dfrac{3}{x} + 2x \quad \dfrac{1}{2} = 2x \quad \dfrac{5}{x}$ $\qquad$ Multiply both sides by 2x

$\qquad\qquad 6 + x = 10$
$\qquad\qquad\quad x = 4$ $\qquad$ Add -6 to both sides

45. $\qquad\qquad 1 - \dfrac{7}{x} = \dfrac{-6}{x^2}$ $\qquad$ LCD $= x^2$

$x^2(1) + x^2\left(-\dfrac{7}{x}\right) = x^2\left(\dfrac{-6}{x^2}\right)$ $\qquad$ Multiply both sides by x^2

$\qquad\qquad x^2 - 7x = -6$
$\qquad\quad x^2 - 7x + 6 = 0$ $\qquad$ Add 6 to each side
$\qquad (x - 1)(x - 6) = 0$

$x - 1 = 0 \quad$ or $\quad x - 6 = 0$
$\quad x = 1 \qquad\qquad\quad x = 6$

The solutions are 1 and 6.

49. $\qquad\qquad\qquad \dfrac{2}{y^2 - 16} = \dfrac{10}{y^2 + 4y}$

$\qquad\qquad\qquad \dfrac{2}{(y + 4)(y - 4)} = \dfrac{10}{y(y + 4)}$ $\qquad$ LCD $=$
$\qquad\qquad\qquad\qquad\qquad\qquad\qquad\qquad\qquad\qquad y(y + 4)(y - 4)$

$y(y + 4)(y - 4) \cdot \dfrac{2}{(y + 4)(y - 4)} = y(y + 4)(y - 4) \cdot \dfrac{10}{y(y + 4)}$ Multiply both sides by $y(y + 4)(y - 4)$

$\qquad\qquad\qquad\qquad 2y = 1\text{-}Y - 4)$
$\qquad\qquad\qquad\qquad 2y = 10y - 40$ $\qquad$ Add -2y, 40 to both sides

$\qquad\qquad\qquad\qquad 40 = 8y$
$\qquad\qquad\qquad\quad\; 5 = y$ $\qquad$ Divide both sides by 8

53.

	d	r	t
upstream	48	x - 3	$\dfrac{48}{x - 3}$
downstream	72	x + 3	$\dfrac{72}{x + 3}$

time (downstream) = time (upstream)

$$\frac{48}{x - 3} = \frac{72}{x + 3} \qquad\qquad \text{LCD} = (x - 3)(x + 3)$$

$$(x + 3)(x - 3) \cdot \frac{48}{x - 3} = (x + 3)(x - 3) \cdot \frac{72}{x + 3}$$

$$(x + 3)48 = (x - 3)72$$
$$48x + 144 = 72x - 216$$
$$360 = 24x$$
$$15 = x$$

The speed of the boat in still water is 15 mph.

57. $y = \dfrac{4}{x}$

x	y
1	4
2	2
4	1
-1	-4
-2	-2
-4	-1

See the graph on page A-29 in the textbook.

61. $\dfrac{1 - \dfrac{9}{y^2}}{1 + \dfrac{4}{y} - \dfrac{21}{y^2}} = \dfrac{y^2\left(1 - \dfrac{9}{y^2}\right)}{y^2\left(1 + \dfrac{4}{y} - \dfrac{21}{y^2}\right)}$ Multiply top and bottom by LCD = y^2

$= \dfrac{y^2 \cdot 1 - y^2 \cdot \dfrac{9}{y^2}}{y^2 \cdot 1 + y^2 \cdot \dfrac{4}{y} - y^2 \cdot \dfrac{21}{y^2}}$ Distributive property

$= \dfrac{y^2 - 9}{y^2 + 4y - 21}$ Simplify

$= \dfrac{(y + 3)(y - 3)}{(y + 7)(y - 3)}$ Factor

$= \dfrac{y + 3}{y + 7}$ Reduce

65. 40 to 100 $\dfrac{40}{100} = \dfrac{2}{5}$

69. $\dfrac{x}{9} = \dfrac{4}{3}$

$3x = 36$ Product of extremes - product of means
$x = 12$

73. $\dfrac{8}{x - 2} = \dfrac{x}{6}$

$48 = x(x - 2)$ Product of extremes = product of means

$48 = x^2 - 2x$ Distributive property

$0 = x^2 - 2x - 48$ Standard form for a quadratic equation

$0 = (x + 6)(x - 8)$ Factor

$x + 6 = 0$ or $x - 8 = 0$ Set factors equal to 0
$x = -6$ $x = 8$

The solutions are -6 and 8.

Chapter 6 Test

1. $\dfrac{x^2 - 16}{x^2 - 8x + 16} = \dfrac{(x + 4)(x - 4)}{(x - 4)(x - 4)}$

$\qquad\qquad = \dfrac{x + 4}{x - 4}$

2. $\dfrac{10a + 20}{5a^2 + 20a + 20} = \dfrac{\overset{2}{\cancel{10}}(a + 2)}{\underset{1}{\cancel{5}}(a + 2)(a + 2)}$

$\qquad\qquad\qquad = \dfrac{2}{a + 2}$

3. $\dfrac{xy + 7x + 5y + 35}{x^2 + ax + 5x + 5a} = \dfrac{(xy + 7x) + (5y + 35)}{(x^2 + ax) + (5x + 5a)}$

$\qquad\qquad\qquad = \dfrac{x(y + 7) + 5(y + 7)}{x(x + a) + 5(x + a)}$

$\qquad\qquad\qquad = \dfrac{(x + 5)(y + 7)}{(x + 5)(x + a)}$

$\qquad\qquad\qquad = \dfrac{x + 7}{x + a}$

4. $\dfrac{3x - 12}{4} \cdot \dfrac{8}{2x - 8}$

$\qquad = \dfrac{3(x - 4)}{4} \cdot \dfrac{8}{2(x - 4)}$

$\qquad = \dfrac{24(x - 4)}{8(x - 4)}$

$\qquad = 3$

5. $\dfrac{x^2 - 49}{x + 1} \div \dfrac{x + 7}{x^2 - 1}$

$\qquad = \dfrac{(x + 7)(x - 7)}{x + 1} \cdot \dfrac{(x + 1)(x - 1)}{x + 7}$

$\qquad = (x - 7)(x - 1)$

6. $\dfrac{x^2 - 3x - 10}{x^2 - 8x + 15} \div \dfrac{3x^2 + 2x - 8}{x^2 + x - 12}$

$\qquad = \dfrac{x^2 - 3x - 10}{x^2 - 8x + 15} \cdot \dfrac{x^2 + x - 12}{3x^2 + 2x - 8}$

$\qquad = \dfrac{(x - 5)(x + 2)}{(x - 3)(x - 5)} \cdot \dfrac{(x + 4)(x - 3)}{(x + 2)(3x - 4)}$

$\qquad = \dfrac{x + 4}{3x - 4}$

7. $(x^2 - 9)(\dfrac{x + 2}{x + 3})$

$\qquad = \dfrac{(x + 3)(x - 3)}{1}(\dfrac{x + 2}{x + 3})$

$\qquad = (x - 3)(x + 2)$

8. $\dfrac{3}{x - 2} - \dfrac{6}{x - 2} = \dfrac{3 - 6}{x - 2}$

$\qquad\qquad\qquad = \dfrac{-3}{x - 2}$

9. $\dfrac{x}{x^2 - 9} + \dfrac{4}{4x - 12}$

$\qquad = \dfrac{x}{(x + 3)(x - 3)} + \dfrac{4}{4(x - 3)}$

$\qquad = \dfrac{x}{(x + 3)(x - 3)} + \dfrac{1}{x - 3}$

$\qquad = \dfrac{x}{(x + 3)(x - 3)} + \dfrac{1(x + 3)}{(x - 3)(x + 3)}$

$\qquad = \dfrac{2x = 3}{(x + 3)(x - 3)}$

10. $\dfrac{2x}{x^2 - 1} + \dfrac{x}{x^2 - 3x + 2}$

$\qquad = \dfrac{2x}{(x + 1)(x - 1)} + \dfrac{x}{(x - 1)(x - 2)}$

$\qquad = \dfrac{2x(x - 2)}{(x + 1)(x - 1)(x - 2)} + \dfrac{x(x + 1)}{(x - 1)(x - 2)(x + 1)}$

$\qquad = \dfrac{2x^2 - 4x + x^2 = x}{(x + 1)(x - 1)(x - 2)}$

$\qquad = \dfrac{3x^2 - 3x}{(x + 1)(x - 1)(x - 2)}$

$\qquad = \dfrac{3x(x - 1)}{(x + 1)(x - 1)(x - 2)}$

$\qquad = \dfrac{3x}{(x + 1)(x - 2)}$

11.
$$\frac{7}{5} = \frac{x + 2}{3}$$

$$15\left(\frac{7}{5}\right) = \left(\frac{x + 2}{3}\right)15 \qquad \text{LCD} = 15$$

$$21 = (x + 2)5$$
$$21 = 5x + 10$$
$$11 = 5x$$

$$\frac{11}{5} = x$$

12.
$$\frac{10}{x + 4} = \frac{6}{x} - \frac{4}{x}$$

$$x(x + 4)\left(\frac{10}{x + 4}\right) = x(x + 4)\left(\frac{6}{x}\right) - x(x + 4)\left(\frac{4}{x}\right) \qquad \text{LCD} = x(x + 4)$$

$$10x = 6(x + 4) - 4(x + 4)$$
$$10x = 6x + 24 - 4x - 16$$
$$10x = 2x + 8$$
$$8x = 8$$
$$x = 1$$

13.
$$\frac{3}{x - 2} - \frac{4}{x + 1} = \frac{5}{x^2 - x - 2}$$

$$(x + 1)(x - 2)\left(\frac{3}{x - 2}\right) - (x + 1)(x - 2)\left(\frac{4}{x + 1}\right) =$$

$$(x + 1)(x - 2)\left(\frac{5}{(x + 1)(x - 2)}\right) \qquad \text{LCD} = (x + 1)(x - 2)$$

$$3(x + 1) - 4(x - 2) = 5$$
$$3x + 3 - 4x + 8 = 5$$
$$-x + 11 = 5$$
$$-x = -6$$
$$x = 6$$

14.

	d	r	t
upstream	26	x - 2	$\frac{26}{x - 2}$
downstream	34	x + 2	$\frac{34}{x + 2}$

(x = speed of the boat)

Remember d = rt so

$$\frac{d}{r} = t$$

time (downstream) = time (upstream)

$$\frac{26}{x - 2} = \frac{34}{x + 2}$$

$$(x + 2)(x - 2)(\frac{26}{x - 2}) = (x + 2)(x - 2)(\frac{34}{x + 2})$$

$$26(x + 2) = 34(x - 2)$$
$$26x + 52 = 34x - 68$$
$$52 = 8x - 68$$
$$120 = 8x$$
$$15 = x$$

The speed of the boat in still water is 15 mph.

15. Let x = amount of time to empty the pool with both pipes open.

$$\frac{1}{12} \qquad - \qquad \frac{1}{15} \qquad = \qquad \frac{1}{x}$$

Amount of water let in by inlet pipe	Amount of water let out by outlet pipe	Total amount of water in pool

$$60x(\frac{1}{12}) - 60x(\frac{1}{15}) = (\frac{1}{x})60x \qquad \text{Find LCD}$$

$$5x - 4x = 60$$
$$x = 60 \text{ hours}$$

Remember to subtract inlet pipe from the outlet pipe because the pool is full and you are trying to empty the pool.

16. $y = \dfrac{8}{x}$

x	y
1	8
2	4
4	2
8	1
-1	-8
-2	-4
-4	-2
-8	-1

See the graph on page A-29 in the textbook.

17. $\dfrac{\text{solution of alcohol}}{\text{solution of water}}$ $\qquad \dfrac{27}{54} = \dfrac{1}{2}$

total volume $\qquad\qquad 27 + 54 = 81$

$\dfrac{\text{alcohol}}{\text{total volume}}$ $\qquad\qquad \dfrac{27}{81} = \dfrac{1}{3}$

18. $\dfrac{\text{defective parts}}{\text{parts produced}}$ $\qquad \dfrac{8}{100} = \dfrac{x}{1650}$

$8(1650) = 100x \qquad$ Product of extremes =
$\qquad\qquad\qquad\qquad\qquad$ product of means

$13200 = 100x$

$132 = x$

When the machine produces 1,650 parts, 132 will be defective.

19. $\dfrac{1 + \dfrac{1}{x}}{1 - \dfrac{1}{x}} = \dfrac{x\left(1 + \dfrac{1}{x}\right)}{x\left(1 - \dfrac{1}{x}\right)}$

$\qquad = \dfrac{x(1) + x\left(\dfrac{1}{x}\right)}{x(1) - x\left(\dfrac{1}{x}\right)}$

$\qquad = \dfrac{x + 1}{x - 1}$

20. $\dfrac{1 - \dfrac{16}{x^2}}{1 - \dfrac{2}{x} - \dfrac{8}{x^2}} = \dfrac{x^2\left(1 - \dfrac{16}{x^2}\right)}{x^2\left(1 - \dfrac{2}{x} - \dfrac{8}{x^2}\right)}$

$$= \frac{x^2(1) - x^2\left(\dfrac{16}{x^2}\right)}{x^2(1) - x^2\left(\dfrac{2}{x}\right) - x^2\left(\dfrac{8}{x^2}\right)}$$

$$= \frac{x^2 - 16}{x^2 - 2x - 8}$$

$$= \frac{(x + 4)(x - 4)}{(x - 4)(x + 2)}$$

$$= \frac{x + 4}{x + 2}$$

Chapter 7

1. $\sqrt{9} = 3$ Because $3^2 = 9$.

5. $\sqrt{-25}$ Not a real number since there is no real number whose square is -25.

9. $\sqrt{625} = 25$ Because $(25)^2 = 625$.

13. $-\sqrt{64} = -8$ Because -8 is the negative number we square to get 64.

17. $\sqrt{1225} = 35$ Because $(35)^2 = 1225$.

21. $\sqrt[3]{-8} = -2$ Because $(-2)^3 = -8$.

25. $\sqrt[3]{-1} = -1$ Because $(-1)^3 = -1$

29. $-\sqrt[4]{16} = -2$ Because -2 is the negative number we raise to the fourth power to get 16.

33. $\sqrt{9x^2} = 3x$ Because $(3x)^2 = 9x^2$

37. $\sqrt{(a + b)^2} = a + b$

41. $\sqrt[3]{x^3} = x$

45. $\sqrt{x^4} = x^2$ Because $(x^2)^2 = x^4$

49. $\sqrt{25a^8b^4} = 5a^4b^2$ Because $(5a^4b^2)^2 = 25a^8b^4$

53. $\sqrt[3]{27a^{12}} = 3a^4$ Because $(3a^4)^3 = 27a^{12}$

57. $\sqrt{9} + \sqrt{16} = 3 + 4$
$$= 7$$

61. $\sqrt{144} + \sqrt{25} = 12 + 5$
$$= 17$$

65. $\sqrt{x^2 + 6x + 9} = \sqrt{(x + 3)^2}$
$$= x + 3$$

Section 7.1 continued

69. A = \$65   P = \$50

$$r = \frac{\sqrt{A} - \sqrt{P}}{\sqrt{P}}$$

$$r = \frac{\sqrt{65} - \sqrt{50}}{\sqrt{50}}$$

$$r = \frac{8.062 - 7.071}{7.071}$$

$$r = \frac{.991}{7.071}$$

$$r \approx .14 = 14\%$$

73. $\dfrac{x^2 - 16}{x + 4} = \dfrac{(x + 4)(x - 4)}{x + 4}$

$$= x - 4$$

77. $\dfrac{2x^2 - 5x - 3}{x^2 - 3x} = \dfrac{(2x + 1)(x - 3)}{x(x - 3)}$

$$= \frac{2x + 1}{x}$$

Section 7.2

1. $\sqrt{8} = \sqrt{4 \cdot 2}$

$$= \sqrt{4}\ \sqrt{2}$$

$$= 2\sqrt{2} \qquad \text{Since } \sqrt{4} = 2$$

5. $\sqrt[3]{24} = \sqrt[3]{8 \cdot 3}$

$$= \sqrt[3]{8}\ \sqrt[3]{2}$$

$$= 2\ \sqrt[3]{2} \qquad \text{Since } \sqrt[3]{8} = 2$$

9. $\sqrt{45a^2b^2} = \sqrt{9a^2b^2 \cdot 5}$

$$= \sqrt{9a^2b^2}\ \sqrt{5}$$

$$= 3ab\ \sqrt{5} \qquad \text{Since } \sqrt{9a^2b^2} = 3ab$$

13. $\sqrt{32x^4} = \sqrt{16x^4 \cdot 2}$

$$= \sqrt{16x^4}\ \sqrt{2}$$

$$= 4x^2\ \sqrt{2} \qquad \text{Since } \sqrt{16x^4} = 4x^2$$

17. $\frac{1}{2}\sqrt{28x^3} = \frac{1}{2}\sqrt{4x^2 \cdot 7x}$

$\qquad = \frac{1}{2}\sqrt{4x^2}\ \sqrt{7x}$

$\qquad = \frac{1}{2} \cdot 2x\sqrt{7x}$ $\qquad$ Since $\sqrt{4x^2} = 2x$

$\qquad = x\sqrt{7x}$

21. $2a\ \sqrt[3]{27a^3} = 2a\ \sqrt[3]{27a^3 \cdot a^2}$

$\qquad = 2a\ \sqrt[3]{27a^3}\ \sqrt[3]{a^2}$ $\qquad$ Property 1

$\qquad = 2a \cdot 3a\ \sqrt[3]{a^2}$ $\qquad$ Since $\sqrt[3]{27a^3} = 3a$

$\qquad = 6a^2\ \sqrt[3]{a^2}$

25. $3\sqrt{50xy^2} = 3\sqrt{25y^2 \cdot 2x}$

$\qquad = 3\sqrt{25y^2}\ \sqrt{2x}$ $\qquad$ Property 1

$\qquad = 3(5y)\sqrt{2x}$

$\qquad = 15y\sqrt{2x}$ $\qquad$ Since $\sqrt{25y^2} = 5y$

29. $\sqrt{\dfrac{16}{25}} = \dfrac{\sqrt{16}}{\sqrt{25}}$ $\qquad$ Property 2

$\qquad = \dfrac{4}{5}$ $\qquad$ $\sqrt{16} = 4$ and $\sqrt{25} = 5$

33. $\sqrt[3]{\dfrac{8}{27}} = \dfrac{\sqrt[3]{8}}{\sqrt[3]{27}}$ $\qquad$ Property 2

$\qquad = \dfrac{2}{3}$ $\qquad$ $\sqrt[3]{8} = 2$ and $\sqrt[3]{27} = 3$

37. $\sqrt{\dfrac{100x^2}{25}} = \dfrac{\sqrt{100x^2}}{\sqrt{25}}$ $\qquad$ Property 2

$\qquad = \dfrac{10x}{5}$ $\qquad$ $\sqrt{100x^2} = 10x$ and $\sqrt{25} = 5$

$\qquad = 2x$

41. $\sqrt[3]{\dfrac{27x^3}{8y^3}} = \dfrac{\sqrt[3]{27x^3}}{\sqrt[3]{8y^3}}$ $\qquad$ Property 2

$\qquad = \dfrac{3x}{2y}$ $\qquad$ $\sqrt[3]{27x^3} = 3x$ and $\sqrt[3]{8y^3} = 2y$

45. $\sqrt{\dfrac{75}{25}} = \sqrt{3}$

$\dfrac{75}{25} = 3$

49. $\sqrt{\dfrac{288x}{25}} = \dfrac{\sqrt{288x}}{\sqrt{25}}$　　　Property 2

$\qquad\quad = \dfrac{\sqrt{144 \cdot 2x}}{\sqrt{25}}$

$\qquad\quad = \dfrac{\sqrt{144}\ \sqrt{2x}}{\sqrt{25}}$　　　Property 1

$\qquad\quad = \dfrac{12\sqrt{2x}}{5}$

53. $\dfrac{3\sqrt{50}}{2} = \dfrac{3\sqrt{25}\ \sqrt{2}}{2}$　　　Property 1

$\qquad\quad = \dfrac{3(5)\sqrt{2}}{2}$

$\qquad\quad = \dfrac{15\sqrt{2}}{2}$

57. $\dfrac{5\sqrt{72a^2b^2}}{\sqrt{36}} = \dfrac{5\sqrt{36a^2b^2 \cdot 2}}{\sqrt{36}}$

$\qquad\quad = \dfrac{5\sqrt{36a^2b^2} \cdot \sqrt{2}}{\sqrt{36}}$　　　Property 1

$\qquad\quad = \dfrac{5(6)ab\sqrt{2}}{6}$

$\qquad\quad = 5ab\sqrt{2}$

61. $\dfrac{8\sqrt{12x^2y^3}}{\sqrt{100}} = \dfrac{8\sqrt{4x^2y^2 \cdot 3y}}{\sqrt{100}}$

$\qquad\quad = \dfrac{8\sqrt{4x^2y^2}\ \sqrt{3y}}{\sqrt{100}}$　　　Property 1

$\qquad\quad = \dfrac{8(2)xy\sqrt{3y}}{10}$

$\qquad\quad = \dfrac{8xy\sqrt{3y}}{5}$

Section 7.2 continued

65. Let h = 25 feet, then

$$t = \sqrt{\frac{h}{16}} \qquad \text{becomes}$$

$$t = \sqrt{\frac{25}{16}}$$

$$= \frac{5}{4} \text{ seconds}$$

69. $\dfrac{x^2 + 3x - 4}{3x^2 + 7x - 20} \div \dfrac{x^2 - 2x + 1}{3x^2 - 2x - 5}$

$$= \frac{x^2 + 3x - 4}{3x^2 + 7x - 20} \cdot \frac{3x^2 - 2x - 5}{x^2 - 2x + 1}$$

$$= \frac{(x + 4)(x - 1)(x + 1)(3x - 5)}{(x + 4)(3x - 5)(x - 1)(x - 1)}$$

$$= \frac{x + 1}{x - 1}$$

Section 7.3

1. $\sqrt{\dfrac{1}{2}} = \dfrac{\sqrt{1}}{\sqrt{2}}$ 　　　　　Property 2

$$= \frac{1}{\sqrt{2}} \cdot \frac{\sqrt{2}}{\sqrt{2}}$$

$$= \frac{\sqrt{2}}{2} \qquad\qquad \sqrt{2} \cdot \sqrt{2} = \sqrt{4} = 2$$

5. $\sqrt{\dfrac{2}{5}} = \dfrac{\sqrt{2}}{\sqrt{5}}$ 　　　　　Property 2

$$= \frac{\sqrt{2}}{5} \cdot \frac{\sqrt{5}}{\sqrt{5}}$$

$$= \frac{\sqrt{10}}{5} \qquad\qquad \sqrt{5} \cdot \sqrt{5} = \sqrt{25} = 5$$

9. $\sqrt{\dfrac{20}{3}} = \dfrac{\sqrt{4}\,\sqrt{5}}{\sqrt{3}}$ 　　　　　Property 1

$$= \frac{2\sqrt{5}}{3} \cdot \frac{\sqrt{3}}{\sqrt{3}}$$

$$= \frac{2\sqrt{15}}{3} \qquad\qquad \sqrt{3} \cdot \sqrt{3} = \sqrt{9} = 3$$

Section 7.3 continued

13. $\sqrt{\dfrac{20}{5}} = \sqrt{4}$ $\qquad \dfrac{20}{5}$

$\qquad\quad = 2$

17. $\dfrac{\sqrt{35}}{\sqrt{7}} = \sqrt{\dfrac{35}{7}}$ $\qquad$ Property 2

$\qquad\quad = \sqrt{5}$ $\qquad \dfrac{35}{7} = 5$

21. $\dfrac{6\sqrt{21}}{3\sqrt{7}} = \dfrac{6}{3}\sqrt{\dfrac{21}{7}}$

$\qquad\quad = 2\sqrt{3}$ $\qquad \dfrac{21}{7} = 3$

25. $\sqrt{\dfrac{4x^2y^2}{2}} = \sqrt{2x^2y^2}$

$\qquad\quad = xy\sqrt{2}$ $\qquad$ Since $\sqrt{x^2y^2} = xy$

29. $\sqrt{\dfrac{16a^4}{5}} = \dfrac{\sqrt{16a^4}}{\sqrt{5}}$

$\qquad\quad = \dfrac{4a}{\sqrt{5}} \cdot \dfrac{\sqrt{5}}{\sqrt{5}}$

$\qquad\quad = \dfrac{4a^2\sqrt{5}}{5}$

33. $\sqrt{\dfrac{20x^2y^3}{3}} = \dfrac{\sqrt{4x^2y^3 \cdot 5y}}{\sqrt{3}}$

$\qquad\quad = \dfrac{2xy\sqrt{5y}}{\sqrt{3}} \cdot \dfrac{\sqrt{3}}{\sqrt{3}}$

$\qquad\quad = \dfrac{2xy\sqrt{15y}}{3}$

37. $\dfrac{6\sqrt{54a^2b^3}}{5} = \dfrac{6\sqrt{9a^2b^2 \cdot 6b}}{5}$

$\qquad\quad = \dfrac{6(3)ab\sqrt{6b}}{5}$

$\qquad\quad = \dfrac{18ab\sqrt{6b}}{5}$

Section 7.3 continued

41. $\sqrt[3]{\dfrac{1}{2}} = \dfrac{\sqrt[3]{1}}{\sqrt[3]{2}}$

$\qquad = \dfrac{1}{\sqrt[3]{2}} \cdot \dfrac{\sqrt[3]{4}}{\sqrt[3]{4}}$

$\qquad = \dfrac{\sqrt[3]{4}}{\sqrt[3]{8}}$

$\qquad = \dfrac{\sqrt[3]{4}}{2}$

45. $\sqrt[3]{\dfrac{3}{2}} = \dfrac{\sqrt[3]{3}}{\sqrt[3]{2}} \cdot \dfrac{\sqrt[3]{4}}{\sqrt[3]{4}}$

$\qquad = \dfrac{\sqrt[3]{12}}{\sqrt[3]{8}}$

$\qquad = \dfrac{\sqrt[3]{12}}{2}$

49. $\dfrac{1}{\sqrt{2}} = \dfrac{1}{1.414}$

$\qquad = .707$

$\dfrac{\sqrt{2}}{2} = \dfrac{1.414}{2}$ $\qquad$ Both $\dfrac{1}{\sqrt{2}}$ and $\dfrac{\sqrt{2}}{2}$ equal .707.

$\qquad = .707$

53. $3x + 7x = (3 + 7)x = 10x$

57. $7a - 3a + 6a = (7 - 3 + 6)a = 10a$

61. $\dfrac{a}{3} + \dfrac{2}{5}$ $\qquad\qquad$ LCD is 15

$\qquad = \dfrac{a}{3} \cdot \dfrac{5}{5} + \dfrac{2}{5} \cdot \dfrac{3}{3}$

$\qquad = \dfrac{5a}{15} + \dfrac{6}{15}$

$\qquad = \dfrac{5a + 6}{15}$

Section 7.4

1. $3\sqrt{2} + 4\sqrt{2}$

$\qquad = (3 + 4)\sqrt{2}$ $\qquad$ Distributive property

$\qquad = 7\sqrt{2}$

Section 7.4 continued

5. $\sqrt{3} + 6\sqrt{3}$

   $= 1\sqrt{3} + 6\sqrt{3}$

   $= (1 + 6)\sqrt{3}$ Distributive property

   $= 7\sqrt{3}$

9. $14\sqrt{13} - \sqrt{13}$

   $= 14\sqrt{13} - 1\sqrt{13}$

   $= (14 - 1)\sqrt{13}$

   $= 13\sqrt{13}$

13. $5\sqrt{5} + \sqrt{5}$

   $= 5\sqrt{5} + 1\sqrt{5}$

   $= (5 + 1)\sqrt{5}$ Distributive property

   $= 6\sqrt{5}$

17. $3\sqrt{3} = \sqrt{27}$

   $= 3\sqrt{3} - \sqrt{9}\,\sqrt{3}$

   $= 3\sqrt{3} - 3\sqrt{3}$

   $= (3 - 3)\sqrt{3}$

   $= 0$

21. $-\sqrt{75} - \sqrt{3}$

   $= -\sqrt{25}\,\sqrt{3} - \sqrt{3}$

   $= -5\sqrt{3} - 1\sqrt{3}$

   $= (-5 - 1)\sqrt{3}$

   $= -6\sqrt{3}$

25. $\frac{3}{4}\sqrt{8} + \frac{3}{10}\sqrt{75}$

   $= \frac{3}{4} \cdot 8 \sqrt{4}\,\sqrt{2} + \frac{3}{10}\sqrt{25}\,\sqrt{3}$

   $= \frac{3}{\cancel{4}} \cdot \cancel{8}^{\,2} \cdot 2\sqrt{2} + \frac{3}{\cancel{10}_{\,2}} \cdot 5\sqrt{3}$

   $= 12\sqrt{2} + \frac{3}{2}\sqrt{3}$

29. $\frac{5}{6}\sqrt{72} - \frac{3}{8}\sqrt{8} + \frac{3}{10}\sqrt{50}$

$= \frac{5}{6}\sqrt{36}\sqrt{2} - \frac{3}{8}\sqrt{4}\sqrt{2} + \frac{3}{10}\sqrt{25}\sqrt{2}$

$= \frac{5}{6} \cdot 6\sqrt{2} - \frac{3}{8} \cdot 2\sqrt{2} + \frac{3}{10} \cdot 5\sqrt{2}$

$= 5\sqrt{2} - \frac{3}{4}\sqrt{2} + \frac{3}{2}\sqrt{2} = (\frac{20}{4} - \frac{3}{4} + \frac{6}{4})\sqrt{20} = \frac{23}{4}\sqrt{2}$

33. $6\sqrt{48} - 2\sqrt{12} + 5\sqrt{27}$

$= 6\sqrt{16}\sqrt{3} - 2\sqrt{4}\sqrt{3} + 5\sqrt{9}\sqrt{3}$

$= 6 \cdot 4\sqrt{3} - 2 \cdot 2\sqrt{3} + 5 \cdot 3\sqrt{3}$

$= 24\sqrt{3} - 4\sqrt{3} + 15\sqrt{3}$

$= (24 - 4 + 15)\sqrt{3}$

$= 35\sqrt{3}$

37. $\sqrt{x^3} + x\sqrt{x}$

$= \sqrt{x^2}\sqrt{x} + x\sqrt{x}$

$= x\sqrt{x} + x\sqrt{x}$

$= (x + x)\sqrt{x}$

$= 2x\sqrt{x}$

41. $5\sqrt{8x^3} + x\sqrt{50x}$

$= 5\sqrt{4x^2}\sqrt{2x} + x\sqrt{25}\sqrt{2x}$

$= 5 \cdot 2x\sqrt{2x} + x \cdot 5\sqrt{2x}$

$= 10x\sqrt{2x} + 5x\sqrt{2x}$

$= (10x + 5x)\sqrt{2x}$

$= 15x\sqrt{2x}$

45. $\sqrt{20ab^2} - b\sqrt{45a}$

$= \sqrt{4b^2}\sqrt{5a} - b\sqrt{9}\sqrt{5a}$

$= 2b\sqrt{5a} - b \cdot 3\sqrt{5a}$

$= (2b - 3b)\sqrt{5a}$

$= -b\sqrt{5a}$

Section 7.4 continued

49. $7\sqrt{50x^2y} + 8x\sqrt{8y} - 7\sqrt{32x^2y}$

$$= 7\sqrt{25x^2}\,\sqrt{2y} + 8x\sqrt{4}\,\sqrt{2y} - 7\sqrt{16x^2}\,\sqrt{2y}$$

$$= 7 \cdot 5x\sqrt{2y} + 8x \cdot 2\sqrt{2y} - 7 \cdot 4x\sqrt{2y}$$

$$= 35x\sqrt{2y} + 16x\sqrt{2y} - 28x\sqrt{2y}$$

$$= (35x + 16x - 28x)\sqrt{2y}$$

$$= 23x\sqrt{2y}$$

53. $4\sqrt{3} + 5\sqrt{3} = 9\sqrt{6}$ Given statement is false

$\quad\,\, 4\sqrt{3} + 5\sqrt{3} = 9\sqrt{3}$ Corrected statement

57. $(3x - 4y)(3x + 4y)$

$\quad = 9x^2 + 12xy - 12xy - 16y^2$ FOIL method

$\quad = 9x^2 - 16y^2$

61.
$$1 - \frac{5}{x} = \frac{-6}{x^2}$$

$$x^2(1) - x^2\left(\frac{5}{x}\right) = \left(\frac{-6}{x^2}\right)x^2 \qquad \text{Multiply by } x^2$$

$$x^2 - 5x = -6$$
$$x^2 - 5x + 6 = 0$$
$$(x - 2)(x - 3) = 0$$

$x - 2 = 0 \quad \text{or} \quad x - 3 = 0$

$\qquad x = 2 \qquad\qquad x = 3$

The solutions are 2 and 3.

Section 7.5

1. $\sqrt{3}\,\sqrt{2} = \sqrt{3 \cdot 2} = \sqrt{6}$

5. $(2\sqrt{3})(5\sqrt{7})$

$\quad = (2 \cdot 5)(\sqrt{3}\,\sqrt{7})$ Commutative and associative properties

$\quad = (2 \cdot 5)(\sqrt{3 \cdot 7})$ Property 1 of radicals

$\quad = 10\sqrt{21}$ Multiply

9. $\sqrt{2}(\sqrt{3} - 1)$

$\quad = \sqrt{2} \cdot \sqrt{3} - \sqrt{2}(1)$ Distributive property

$\quad = \sqrt{6} - \sqrt{2}$ Multiply

Section 7.5 continued

13. $\sqrt{3}\,(2\sqrt{2} + \sqrt{3}\,)$

$\qquad = \sqrt{3} \cdot 2\sqrt{2} + \sqrt{3} \cdot \sqrt{3} \qquad$ Distributive property

$\qquad = 2\sqrt{6} + 3 \qquad\qquad\qquad$ Multiply

17. $2\sqrt{3}\,(\sqrt{2} + \sqrt{5}\,)$

$\qquad = 2\sqrt{3} \cdot \sqrt{2} + 2\sqrt{3} \cdot \sqrt{5} \qquad$ Distributive property

$\qquad = 2\sqrt{6} + 2\sqrt{15}$

21. $(\sqrt{x} + 3)^2$

$\qquad = (\sqrt{x} + 3)(\sqrt{x} + 3)$

$\qquad = \underset{F}{\sqrt{x} \cdot \sqrt{x}} + \underset{O}{3\sqrt{x}} + \underset{I}{3\sqrt{x}} + \underset{L}{3 \cdot 3}$

$\qquad = x + 6\sqrt{x} + 9$

25. $\left(\sqrt{a} - \dfrac{1}{2}\right)^2$

$\qquad = \left(\sqrt{a} - \dfrac{1}{2}\right)\left(\sqrt{a} - \dfrac{1}{2}\right)$

$\qquad = \underset{F}{\sqrt{a}\,\sqrt{a}} - \underset{O}{\dfrac{1}{2}\sqrt{a}} - \underset{I}{\dfrac{1}{2}\sqrt{a}} + \underset{L}{\dfrac{1}{4}}$

$\qquad = a - \sqrt{a} + \dfrac{1}{4}$

33. $\left(\sqrt{3} + \dfrac{1}{2}\right)\left(\sqrt{2} + \dfrac{1}{3}\right)$

$\qquad = \underset{F}{\sqrt{3}\,\sqrt{2}} + \underset{O}{\dfrac{1}{3}\sqrt{3}} + \underset{I}{\dfrac{1}{2}\sqrt{2}} + \underset{L}{\dfrac{1}{6}}$

$\qquad = \sqrt{6} + \dfrac{1}{3}\sqrt{3} + \dfrac{1}{2}\sqrt{2} + \dfrac{1}{6}$

37. $\left(\sqrt{a} + \dfrac{1}{3}\right)\left(\sqrt{a} + \dfrac{2}{3}\right)$

$\qquad = \underset{F}{\sqrt{a}\,\sqrt{a}} + \underset{O}{\dfrac{2}{3}\sqrt{a}} + \underset{I}{\dfrac{1}{3}\sqrt{a}} + \underset{L}{\dfrac{2}{9}}$

$\qquad = a + \sqrt{a} + \dfrac{2}{9}$

Section 7.5 continued

41. $(2\sqrt{7} + 3)(3\sqrt{7} - 4)$

$\quad = 2\sqrt{7} \cdot 3\sqrt{7} - 4 \cdot 2\sqrt{7} + 3 \cdot 3\sqrt{7} + 3(-4)$

$\qquad\qquad$ F $\qquad\qquad$ O $\qquad\quad$ I $\qquad$ L

$\quad = 6 \cdot 7 - 8\sqrt{7} + 9\sqrt{7} - 12$

$\quad = 42 + \sqrt{7} - 12$

$\quad = 30 + \sqrt{7}$

45. $(7\sqrt{a} + 2\sqrt{b})(7\sqrt{a} - 2\sqrt{b})$

$\quad = (7\sqrt{a})^2 - (2\sqrt{b})^2$

$\quad = 49a - 4b$

49. $\dfrac{\sqrt{5}}{\sqrt{5} + \sqrt{2}}$

$\quad = \dfrac{\sqrt{5}}{\sqrt{5} + \sqrt{2}} \quad \dfrac{(\sqrt{5} - \sqrt{2})}{(\sqrt{5} - \sqrt{2})}$

$\quad = \dfrac{\sqrt{5}\sqrt{5} - \sqrt{5}\sqrt{2}}{(\sqrt{5})^2 - (\sqrt{2})^2}$

$\quad = \dfrac{5 - \sqrt{10}}{5 - 2}$

$\quad = \dfrac{5 - \sqrt{10}}{3}$

53. $\dfrac{\sqrt{3} + \sqrt{2}}{\sqrt{3} - \sqrt{2}}$

$\quad = \dfrac{\sqrt{3} + \sqrt{2}}{\sqrt{3} - \sqrt{2}} \cdot \dfrac{(\sqrt{3} + \sqrt{2})}{(\sqrt{3} + \sqrt{2})}$

$\quad = \dfrac{\sqrt{3}\sqrt{3} + \sqrt{3}\sqrt{2} + \sqrt{2}\sqrt{3} + \sqrt{2}\sqrt{2}}{(\sqrt{3})^2 - (\sqrt{2})^2}$

$\quad = \dfrac{3 + \sqrt{6} + \sqrt{6} + 2}{3 - 2}$

$\quad = \dfrac{5 + 2\sqrt{6}}{1}$

$\quad = 5 + 2\sqrt{6}$

57. $\dfrac{\sqrt{x} + 2}{\sqrt{x} - 2}$

$\quad = \dfrac{\sqrt{x} + 2}{\sqrt{x} - 2} \cdot \dfrac{(\sqrt{x} + 2)}{(\sqrt{x} + 2)}$

$\quad = \dfrac{\sqrt{x}\sqrt{x} + 2\sqrt{x} + 2\sqrt{x} + 2 \cdot 2}{(\sqrt{x})^2 - 2^2}$

$\quad = \dfrac{x + 4\sqrt{x} + 4}{x - 4}$

Section 7.5 continued

61. $2(2\sqrt{5}) = 6\sqrt{15}$ Given statement is false

$2(3\sqrt{5}) = 6\sqrt{5}$ Corrected statement

65.
$$x^2 + 5x - 6 = 0$$
$$(x + 6)(x - 1) = 0$$

$x + 6 = 0$ or $x - 1 = 0$

$x = -6$ $x = 1$

The solutions are -6 and 1.

69. $\dfrac{x}{3} = \dfrac{27}{x}$

$x \cdot x = 3 \cdot 27$ Product of extremes = product of means

$x^2 = 81$

$x = 9, -9$

The solutions are 9 and -9.

73. $\dfrac{\text{miles}}{\text{hours}} = \dfrac{375}{15} = \dfrac{m}{20}$

$375 \cdot 20 = 15m$ Product of extremes = product of means

$7500 = 15m$

$500 = m$

You would drive 500 miles in 20 hours.

Section 7.6

1. $\sqrt{x + 1} = 2$

$(\sqrt{x + 1})^2 = 2^2$ Square both sides

$x + 1 = 4$

$x = 3$

To check our solution when $x = 3$,

$\sqrt{3 + 1} = 2$

$\sqrt{4} = 2$

$2 = 2$ A true statement.

5. $\sqrt{x - 9} = -6$

$(\sqrt{x - 9})^2 = (-6)^2$ Square both sides

$x - 9 = 36$

$x = 45$

To check our solution when $x = 45$,

$\sqrt{45 - 9} = -6$

$\sqrt{36} = -6$

$6 = -6$ A false statement

The solution is $\emptyset$.

9. $\sqrt{x - 8} = 0$

   $(\sqrt{x - 8})^2 = 0^2$ ⠀⠀Square both sides

   ⠀⠀⠀$x - 8 = 0$

   ⠀⠀⠀⠀⠀⠀$x = 8$

   To check our solution when x = 8,

   ⠀$\sqrt{8 - 8} = 0$

   ⠀⠀⠀$\sqrt{0} = 0$

   ⠀⠀⠀⠀⠀$0 = 0$ ⠀⠀A true statement

13. $\sqrt{2x - 3} = -5$

   $(\sqrt{2x - 3})^2 = (-5)^2$ ⠀⠀Square both sides

   ⠀⠀$2x - 3 = 25$

   ⠀⠀⠀⠀$2x = 28$

   ⠀⠀⠀⠀⠀$x = 14$

   To check our solution when x = 14,

   $\sqrt{2(14) - 3} = -5$

   ⠀$\sqrt{28 - 3} = -5$

   ⠀⠀⠀$\sqrt{25} = -5$

   ⠀⠀⠀⠀⠀$5 = -5$ ⠀⠀A false statement

   The solution is Ø.

17. ⠀$2\sqrt{x} = 10$

   ⠀⠀$\sqrt{x} = 5$ ⠀⠀Divide by 2

   $(\sqrt{x})^2 = 5^2$ ⠀⠀Square both sides

   ⠀⠀⠀$x = 25$

   To check our solution when x = 25,

   $2\sqrt{25} = 10$

   $2 \cdot 5 = 10$

   ⠀⠀$10 = 10$ ⠀⠀A true statement

21. ⠀$\sqrt{3x + 4} - 3 = 2$

   ⠀⠀$\sqrt{3x + 4} = 5$ ⠀⠀Add 3 to both sides

   $(\sqrt{3x + 4})^2 = 5^2$ ⠀⠀Square both sides

   ⠀⠀$3x + 4 = 25$

   ⠀⠀⠀⠀$3x = 21$

   ⠀⠀⠀⠀⠀$x = 7$

   To check our solution when x = 7,

   $\sqrt{3(7) + 4} - 3 = 2$

   ⠀$\sqrt{21 + 4} - 3 = 2$

   ⠀⠀⠀$\sqrt{25} - 3 = 2$

   ⠀⠀⠀⠀$5 - 3 = 2$

   ⠀⠀⠀⠀⠀⠀$2 = 2$ ⠀⠀A true statement

25.
$$\sqrt{2x + 1} + 5 = 2$$
$$\sqrt{2x + 1} = -3 \qquad \text{Add -5 to both sides}$$
$$(\sqrt{2x + 1})^2 = (-3)^2 \qquad \text{Square both sides}$$
$$2x + 1 = 9$$
$$2x = 8 \qquad \text{Add -1 to both sides}$$
$$x = 4 \qquad \text{Divide by 2}$$

To check our solution when $x = 4$,
$$\sqrt{2(4) + 1} + 5 = 2$$
$$\sqrt{8 + 1} + 5 = 2$$
$$\sqrt{9} + 5 = 2$$
$$3 + 5 = 2$$
$$8 = 2 \qquad \text{A false statement}$$

The solution is $\emptyset$.

29.
$$\sqrt{a + 2} = a + 2$$
$$(\sqrt{a + 2})^2 = (a + 2)^2 \qquad \text{Square both sides}$$
$$a + 2 = a^2 + 4a + 4$$
$$0 = a^2 + 3a + 2 \qquad \text{Add -(a + 2) to both sides}$$
$$0 = (a + 2)(a + 1) \qquad \text{Factor}$$

$a + 2 = 0 \quad$ or $\quad a + 1 = 0$
$\qquad a = -2 \qquad\qquad a = -1$

Check $a = -2$
$$\sqrt{-2 + 2} = -2 + 2$$
$$0 = 0 \qquad \text{A true statement}$$

Check $a = -1$
$$\sqrt{-1 + 2} = -1 + 2$$
$$\sqrt{1} = 1$$
$$1 = 1 \qquad \text{A true statement}$$

33.
$$\sqrt{y - 4} = y - 6$$
$$(\sqrt{y - 4})^2 = (y - 6)^2 \qquad \text{Square both sides}$$
$$y - 4 = y^2 - 12y + 36$$
$$0 = y^2 - 13y + 40 \qquad \text{Add -(y - 4) to both sides}$$
$$0 = (y - 8)(y - 5) \qquad \text{Factor}$$

$0 = y - 8 \quad$ or $\quad 0 = y - 5$
$8 = y \qquad\qquad\qquad 5 = y$

Check $y = 8$
$$\sqrt{8 - 4} = 8 - 6$$
$$\sqrt{4} = 2$$
$$2 = 2 \qquad \text{A true statement}$$

Check $y = 5$
$$\sqrt{5 - 4} = 5 - 6$$
$$\sqrt{1} = -1$$
$$1 = -1 \qquad \text{A false statement}$$

Possible solutions are 5 and 8 but only 8 checks.

37. Let $\hspace{3cm}$ x = a number

then $\hspace{2.5cm}$ $x + 2 = \sqrt{8x}$

becomes $\hspace{2cm}$ $(x + 2)^2 = (\sqrt{8x})^2$

$\hspace{3.5cm} x^2 + 4x + 4 = 8x$ $\hspace{1cm}$ Square both sides

$\hspace{3.5cm} x^2 - 4x + 4 = 0$ $\hspace{1cm}$ Add -8x to both sides

$\hspace{3.5cm} (x - 2)(x - 2) = 0$ $\hspace{1cm}$ Factor

$\hspace{2.5cm} x - 2 = 0 \hspace{0.3cm} \text{or} \hspace{0.3cm} x - 2 = 0$

$\hspace{3cm} x = 2 \hspace{2cm} x = 2$

Check x = 2

$\hspace{1.5cm} 2 + 2 = \sqrt{8(2)}$

$\hspace{2.5cm} 4 = \sqrt{16}$

$\hspace{2.5cm} 4 = 4 \hspace{2cm}$ A true statement

41. Let $\hspace{2cm}$ T = 2

then $\hspace{1.5cm}$ $T = \dfrac{11}{7}\sqrt{\dfrac{L}{2}}$

becomes $\hspace{1cm}$ $2 = \dfrac{11}{7}\sqrt{\dfrac{L}{2}}$

$\hspace{2cm} 2\left(\dfrac{7}{11}\right) = \dfrac{11}{7}\left(\dfrac{7}{11}\right)\sqrt{\dfrac{L}{2}} \hspace{1cm}$ Multiply by $\dfrac{7}{11}$

$\hspace{3cm} \dfrac{14}{11} = \sqrt{\dfrac{L}{2}}$

$\hspace{2.5cm} \left(\dfrac{14}{11}\right)^2 = \left(\sqrt{\dfrac{L}{2}}\right)^2 \hspace{1cm}$ Square both sides

$\hspace{3cm} \dfrac{196}{121} = \dfrac{L}{2}$

$\hspace{2cm} 2\left(\dfrac{196}{121}\right) = 2\left(\dfrac{L}{2}\right) \hspace{1cm}$ Multiply by 2

$\hspace{3cm} \dfrac{392}{121} = L$

$\hspace{3cm} \dfrac{392}{121} \approx 3.2 \text{ feet}$

45. $\hspace{1cm}$ $\sqrt{x + 8} = \sqrt{x} + 2$

$\hspace{0.5cm} (\sqrt{x + 8})^2 = (\sqrt{x} + 2)^2 \hspace{1cm}$ Square both sides

$\hspace{1.5cm} x + 8 = x + 4\sqrt{x} + 4$

$\hspace{2.5cm} 4 = 4\sqrt{x} \hspace{1cm}$ Add -x and -4 to both sides

$\hspace{1.5cm} (4)^2 = (4\sqrt{x})^2 \hspace{1cm}$ Square both sides

$\hspace{2cm} 16 = 16x$

$\hspace{2.5cm} 1 = x$

49. $y = 2\sqrt{x}$

x	y
0	$2\sqrt{0} = 0$
1	$2\sqrt{1} = 2$
4	$2\sqrt{4} = 4$
9	$2\sqrt{9} = 6$

See the graph on page A-31 in the textbook.

53.

$$\frac{\frac{2}{5}}{\frac{4}{15}} = \frac{2}{5} \div \frac{4}{15}$$

$$= \frac{2}{5} \cdot \frac{15}{4}$$

$$= \frac{30}{20}$$

$$= \frac{3}{2}$$

57. Let x = a number and $\frac{1}{x}$ = its reciprocal, then

$$x + \frac{1}{x} = \frac{10}{3}$$

$$3x(x) + 3x(\frac{1}{x}) = 3x(\frac{10}{3}) \qquad \text{Multiply both sides by LCD = 3x}$$

$$3x^2 + 3 = 10x$$
$$3x^2 - 10x + 3 = 0 \qquad \text{Add -10x to each side}$$
$$(3x - 1)(x - 3) = 0 \qquad \text{Factor}$$
$$3x - 1 = 0 \quad \text{or} \quad x - 3 = 0$$
$$3x = 1 \qquad\qquad x = 3$$
$$x = \frac{1}{3}$$

$$\frac{1}{x} = 3 \qquad\qquad \frac{1}{x} = \frac{1}{3}$$

The solution is 3 and $\frac{1}{3}$.

Section 7.6 continued

61. Let x = the sink filled

Hot water faucet + Cold water faucet = Filled sink

$$\frac{1}{4} \qquad + \qquad \frac{1}{3} \qquad = \qquad \frac{1}{x}$$

$$\frac{1}{4} + \frac{1}{3} = \frac{1}{x}$$

$$12x\left(\frac{1}{4}\right) + 12x\left(\frac{1}{3}\right) = 12x\left(\frac{1}{x}\right) \qquad \text{Multiply both sides by LCD = 12x}$$

$$3x + 4x = 12$$
$$7x = 12$$
$$x = \frac{12}{7}$$

It will take $\frac{12}{7}$ minutes to fill the sink.

Chapter 7 Review

1. $\sqrt{25} = 5$

5. $\sqrt[3]{-1} = -1$ $(-1)^3 = -1$

9. $\sqrt{100x^2y^4} = \sqrt{100} \cdot \sqrt{x^2} \cdot \sqrt{y^4}$
$$= 10xy^2$$

13. $\sqrt{24} = \sqrt{4 \cdot 6}$
$$= \sqrt{4} \cdot \sqrt{6} \qquad \text{Property 1}$$
$$= 2\sqrt{6}$$

17. $\sqrt{90x^3y^4} = \sqrt{9x^2y^4 \cdot 10x}$
$$= \sqrt{9x^2y^4} \cdot \sqrt{10x} \qquad \text{Property 1}$$
$$= 3xy^2\sqrt{10x}$$

21. $3\sqrt{20x^3y} = 3\sqrt{4x^2 \cdot 5xy}$
$$= 3\sqrt{4x^2} \cdot \sqrt{5xy} \qquad \text{Property 1}$$
$$= 3 \cdot 2x\sqrt{5xy}$$
$$= 6x\sqrt{5xy}$$

25. $\sqrt{\dfrac{8}{81}} = \dfrac{\sqrt{8}}{\sqrt{81}} \qquad \text{Property 2}$
$$= \dfrac{\sqrt{4 \cdot 2}}{9} \qquad \sqrt{81} = 9$$
$$= \dfrac{2\sqrt{2}}{9} \qquad \sqrt{4} = 2$$

29. $\sqrt{\dfrac{49a^2b^2}{16}} = \dfrac{\sqrt{49a^2b^2}}{\sqrt{16}} \qquad \text{Property 2}$
$$= \dfrac{7ab}{4}$$

33. $\sqrt{\dfrac{40a^2}{121}} = \dfrac{\sqrt{40a^2}}{\sqrt{121}} \qquad \text{Property 2}$
$$= \dfrac{\sqrt{4a^2 \cdot 10}}{11}$$
$$= \dfrac{2a\sqrt{10}}{11} \qquad \sqrt{4a^2} = 2a$$

37. $\dfrac{3\sqrt{120a^2b^3}}{\sqrt{25}} = \dfrac{3\sqrt{4a^2b^2 \cdot 30}}{25}$

$\qquad\qquad\qquad = \dfrac{3 \cdot 2ab\sqrt{30}}{5}$

$\qquad\qquad\qquad = \dfrac{6ab\sqrt{30}}{5}$

41. $\dfrac{2}{\sqrt{7}} = \dfrac{2}{\sqrt{7}} \cdot \dfrac{7}{\sqrt{7}}$

$\qquad\quad = \dfrac{2\sqrt{7}}{7} \qquad\qquad (\sqrt{7})^2 = 7$

45. $\sqrt{\dfrac{5}{48}} = \dfrac{\sqrt{5}}{\sqrt{48}}$ $\qquad\qquad$ Property 2

$\qquad\quad = \dfrac{\sqrt{5}}{\sqrt{16 \cdot 3}}$

$\qquad\quad = \dfrac{\sqrt{5}}{4\sqrt{3}} \qquad\qquad \sqrt{16} = 4$

$\qquad\quad = \dfrac{\sqrt{5}}{4\sqrt{3}} \cdot \dfrac{\sqrt{3}}{\sqrt{3}}$

$\qquad\quad = \dfrac{\sqrt{15}}{4 \cdot 3}$

$\qquad\quad = \dfrac{\sqrt{15}}{12}$

49. $\sqrt{\dfrac{32ab^2}{3}} = \dfrac{\sqrt{32ab^2}}{\sqrt{3}}$ $\qquad\qquad$ Property 2

$\qquad\qquad = \dfrac{\sqrt{16b^2 \cdot 2a}}{\sqrt{3}}$

$\qquad\qquad = \dfrac{4b\sqrt{2a}}{\sqrt{3}}$

$\qquad\qquad = \dfrac{4b\sqrt{2a}}{\sqrt{3}} \cdot \dfrac{\sqrt{3}}{\sqrt{3}}$

$\qquad\qquad = \dfrac{4b\sqrt{6a}}{3} \qquad\qquad (\sqrt{3})^2 = 3$

53. $\dfrac{3}{\sqrt{3} - 4} = \left(\dfrac{3}{\sqrt{3} - 4}\right)\left(\dfrac{3 + 4}{\sqrt{3} + 4}\right)$ Multiply by the conjugate $\sqrt{3} + 4$

$= \dfrac{3\sqrt{3} + 12}{(\sqrt{3})^2 - 4^2}$

$= \dfrac{3\sqrt{3} + 12}{3 - 16}$

$= \dfrac{3\sqrt{3} + 12}{-13}$

$= \dfrac{-3\sqrt{3} - 12}{13}$ Do not leave a negative in the denominator

57. $\dfrac{3}{\sqrt{5} - \sqrt{2}} = \left(\dfrac{3}{\sqrt{5} - \sqrt{2}}\right)\left(\dfrac{\sqrt{5} + \sqrt{2}}{\sqrt{5} + \sqrt{2}}\right)$ Multiply by the conjugate $\sqrt{5} + \sqrt{2}$

$= \dfrac{3\sqrt{5} + 3\sqrt{2}}{(\sqrt{5})^2 - (\sqrt{2})^2}$

$= \dfrac{3\sqrt{5} + 3\sqrt{2}}{5 - 2}$

$= \dfrac{3\sqrt{5} + 3\sqrt{2}}{3}$

$= \dfrac{3(\sqrt{5} + \sqrt{2})}{3}$

$= \sqrt{5} + \sqrt{2}$

61. $\dfrac{\sqrt{5} - \sqrt{2}}{\sqrt{5} + \sqrt{2}} = \left(\dfrac{\sqrt{5} - \sqrt{2}}{\sqrt{5} + \sqrt{2}}\right)\left(\dfrac{\sqrt{5} - \sqrt{2}}{\sqrt{5} - \sqrt{2}}\right)$ Multiply by the conjugate $\sqrt{5} - \sqrt{2}$

$= \dfrac{5 - 2\sqrt{10} + 2}{(\sqrt{5})^2 - (\sqrt{2})^2}$ FOIL method

$= \dfrac{7 - 2\sqrt{10}}{5 - 2}$

$= \dfrac{7 - 2\sqrt{10}}{3}$

65. $3\sqrt{5} - 7\sqrt{5} = (3 - 7)\sqrt{5}$

$= -4\sqrt{5}$

69. $-2\sqrt{45} - 5\sqrt{80} + 2\sqrt{20}$

$= -2\sqrt{9 \cdot 5} - 5\sqrt{16 \cdot 5} + 2\sqrt{4 \cdot 5}$

$= -2 \cdot 3\sqrt{5} - 5 \cdot 4\sqrt{5} + 2 \cdot 2\sqrt{5}$

$= -6\sqrt{5} - 20\sqrt{5} + 4\sqrt{5}$

$= (-6 - 20 + 4)\sqrt{5}$

$= -22\sqrt{5}$

73. $\sqrt{40a^3b^2} - a\sqrt{90ab^2}$

$= \sqrt{4 \cdot a^2b^2 \cdot 10a} - a\sqrt{9b^2 \cdot 10a}$

$= 2ab\sqrt{10a} - a \cdot 3b\sqrt{10a}$

$= 2ab\sqrt{10a} - 6ab\sqrt{10a}$

$= (2 - 6)ab\sqrt{10a}$

$= -4ab\sqrt{10a}$

77. $4\sqrt{2}\,(\sqrt{3} + \sqrt{5}\,) = 4\sqrt{2}\,(\sqrt{3}\,) + 4\sqrt{2}\,(\sqrt{5}\,)$

$= 4\sqrt{6} + 4\sqrt{10}$

81. $(2\sqrt{5} - 4)(\sqrt{5} + 3)$

$ \quad\quad\quad\quad F \quad\quad\quad O \quad\quad\quad I \quad\quad\quad L$

$= 2\sqrt{5}\,(\sqrt{5}\,) + 2\sqrt{5}\,(3) + (-4)(\sqrt{5}\,) + (-4)(3)$

$= 2 \cdot 5 + 6\sqrt{5} - 4\sqrt{5} - 12$

$= 10 + 2\sqrt{5} - 12$

$= 2\sqrt{5} - 2$

85. $\sqrt{x - 3} = 3$ $\quad\quad$ Square both sides

$(\sqrt{x - 3}\,)^2 = 3^2$

$\phantom{(\sqrt{x - 3}\,)^2} x - 3 = 9$

$\phantom{(\sqrt{x - 3}\,)^2 x} x = 12$

Check: $\sqrt{12 - 3} = 3$

$\phantom{Check: \sqrt{12 - 3}} \sqrt{9} = 3$

$\phantom{Check: \sqrt{12 - 3} \sqrt{9}} 3 = 3$ $\quad\quad$ A true statement

89. $5\sqrt{a} = 20$ $\quad\quad$ Square both sides

$(5\sqrt{a}\,)^2 = 20^2$

$\phantom{(5\sqrt{a}\,)^2} 25a = 400$

$\phantom{(5\sqrt{a}\,)^2 2} a = 16$

Check: $5\sqrt{16} = 20$

$ 5 \cdot 4 = 20$

$ 20 = 20$ $\quad\quad$ A true statement

93.

$$\sqrt{7x + 1} = x + 1$$

$$(\sqrt{7x + 1})^2 = (x + 1)^2 \qquad \text{Square both sides}$$

$$7x + 1 = x^2 + 2x + 1$$

$$0 = x^2 - 5x$$

$$0 = x(x - 5)$$

$$x = 0 \quad \text{or} \quad x - 5 = 0$$

$$x = 5$$

Check: $\sqrt{7 \cdot 0 + 1} = 0 + 1$

$$\sqrt{1} = 1$$

$$1 = 1 \qquad \text{A true statement}$$

Check: $\sqrt{7 \cdot 5 + 1} = 5 + 1$

$$\sqrt{36} = 6$$

$$6 = 6 \qquad \text{A true statement}$$

97. $y = \sqrt{x} + 3$

x	y
0	$\sqrt{0} + 3 = 3$
1	$\sqrt{1} + 3 = 4$
4	$\sqrt{4} + 3 = 5$

See the graph on page A-32 in the textbook.

Chapter 7 Test

1. $\sqrt{16} = 4$

    Because $(4)^2 = 16$

2. $-\sqrt{36} = -6$

    Because -6 is the negative number we square to get 36.

3. The square roots of 49 and 7 and -7.

4. $\sqrt[3]{27} = 3$

    Because $(3)^3 = 27$

5. $\sqrt[3]{-8} = -2$

    Because $(-2)^3 = -8$

6. $-\sqrt[4]{81} = -3$

    Because -3 is the negative number we raise to the fourth power to get 81.

7. $\sqrt{75} = \sqrt{25 \cdot 3}$

    $\phantom{\sqrt{75}} = \sqrt{25}\,\sqrt{3}$

    $\phantom{\sqrt{75}} = 5\sqrt{3}$

8. $\sqrt{32} = \sqrt{16 \cdot 2}$

    $\phantom{\sqrt{32}} = \sqrt{16}\,\sqrt{2}$

    $\phantom{\sqrt{32}} = 4\sqrt{2}$

9. $\sqrt{\dfrac{2}{3}} = \dfrac{\sqrt{2}}{\sqrt{3}}$

    $\phantom{\sqrt{\dfrac{2}{3}}} = \dfrac{\sqrt{2}}{\sqrt{3}} \cdot \dfrac{\sqrt{3}}{\sqrt{3}}$

    $\phantom{\sqrt{\dfrac{2}{3}}} = \dfrac{\sqrt{6}}{3}$

10. $\dfrac{1}{\sqrt[3]{4}} = \dfrac{1}{\sqrt[3]{4}} \cdot \dfrac{\sqrt[3]{2}}{\sqrt[3]{2}}$

    $\phantom{\dfrac{1}{\sqrt[3]{4}}} = \dfrac{\sqrt[3]{2}}{\sqrt[3]{8}}$

    $\phantom{\dfrac{1}{\sqrt[3]{4}}} = \dfrac{\sqrt[3]{2}}{2}$

11. $3\sqrt{50x^2}$

$\qquad = 3\sqrt{25x^2 \cdot 2}$

$\qquad = 3\sqrt{25x^2}\,\sqrt{2}$

$\qquad = 3 \cdot 5x\sqrt{2}$

$\qquad = 15x\sqrt{2}$

12. $\sqrt{\dfrac{12x^2y^3}{5}}$

$\qquad = \dfrac{\sqrt{12x^2y^3}}{\sqrt{5}}$

$\qquad = \dfrac{\sqrt{4x^2y^2 \cdot 3y}}{\sqrt{5}}$

$\qquad = \dfrac{\sqrt{4x^2y^2}\,\sqrt{3y}}{\sqrt{5}}$

$\qquad = \dfrac{2xy\sqrt{3y}}{\sqrt{5}}$

$\qquad = \dfrac{2xy\sqrt{3y}}{\sqrt{5}} \cdot \dfrac{\boldsymbol{\sqrt{5}}}{\boldsymbol{\sqrt{5}}}$

$\qquad = \dfrac{2xy\sqrt{15y}}{5}$

13. $5\sqrt{12} - 2\sqrt{27}$

$\qquad = 5\sqrt{4 \cdot 3} - 2\sqrt{9 \cdot 3}$

$\qquad = 5\sqrt{4}\,\sqrt{3} - 2\sqrt{9}\,\sqrt{3}$

$\qquad = 5(2)\sqrt{3} - 2(3)\sqrt{3}$

$\qquad = 10\sqrt{3} - 6\sqrt{3}$

$\qquad = 4\sqrt{3}$

14. $2x\sqrt{18} + 5\sqrt{2x^2}$

$\qquad = 2x\sqrt{9 \cdot 2} + 5\sqrt{x^2 \cdot 2}$

$\qquad = 2x\sqrt{9}\,\sqrt{2} + 5\sqrt{x^2}\,\sqrt{2}$

$\qquad = 2x(3)\sqrt{2} + 5x\sqrt{2}$

$\qquad = 6x\sqrt{2} + 5x\sqrt{2}$

$\qquad = 11x\sqrt{2}$

15. $\sqrt{3}(\sqrt{5} - 2)$

$\qquad = \sqrt{3}\,\sqrt{5} - \sqrt{3}(2)$

$\qquad = \sqrt{15} - 2\sqrt{3}$

16. $(\sqrt{5} + 7)(\sqrt{5} - 8)$

$\quad = \sqrt{5}\,\sqrt{5} - 8\sqrt{5} \underline{\quad} 7\sqrt{5} - 7(8)$

$\qquad\qquad$ F $\qquad$ O $\qquad$ I $\qquad$ L

$\quad = \quad 5 \quad - 8\sqrt{5} + 7\sqrt{5} - 56$

$\quad = -51 - \sqrt{5}$

17. $(\sqrt{x} + 6)(\sqrt{x} - 6)$

$\quad = \sqrt{x}\,\sqrt{x} - 6(6)$

$\quad = x - 36$

18. $(\sqrt{5} - \sqrt{3})^2$

$\quad = (\sqrt{5} - \sqrt{3})(\sqrt{5} - \sqrt{3})$

$\quad = \sqrt{5}\,\sqrt{5} - 2(\sqrt{5}\,\sqrt{3}) + \sqrt{3}\,\sqrt{3}$

$\quad = 5 - 2\sqrt{15} + 3$

$\quad = 8 - 2\sqrt{15}$

19. $\dfrac{\sqrt{7} - \sqrt{3}}{\sqrt{7} + \sqrt{3}}$

$\quad = \dfrac{\sqrt{7} - \sqrt{3}}{\sqrt{7} + \sqrt{3}} \cdot \dfrac{\sqrt{7} - \sqrt{3}}{\sqrt{7} - \sqrt{3}}$

$\quad = \dfrac{\sqrt{7}\,\sqrt{7} - 2(\sqrt{7}\,\sqrt{3}) + \sqrt{3}\,\sqrt{3}}{\sqrt{7}\,\sqrt{7} - \sqrt{3}\,\sqrt{3}}$

$\quad = \dfrac{7 - 2\sqrt{21} + 3}{7 - 3}$

$\quad = \dfrac{10 - 2\sqrt{21}}{4}$

$\quad = \dfrac{2(5 - \sqrt{21})}{4}$

$\quad = \dfrac{5 - \sqrt{21}}{2}$

20. $\dfrac{\sqrt{x}}{\sqrt{x} + 5}$

$\quad = \dfrac{\sqrt{x}}{\sqrt{x} + 5} \cdot \dfrac{\sqrt{x} - 5}{\sqrt{x} - 5}$

$\quad = \dfrac{\sqrt{x}\,\sqrt{x} - 5\sqrt{x}}{\sqrt{x}\,\sqrt{x} - 25}$

$\quad = \dfrac{x - 5\sqrt{x}}{x - 25}$

21. $\sqrt{2x + 1} + 2 = 7$

       $\sqrt{2x + 1} = 5$

    $(\sqrt{2x + 1})^2 = 5^2$ Square both sides

        $2x + 1 = 25$

           $2x = 24$

            $x = 12$

To check our solution when x = 12,

    $\sqrt{2(12) + 1} + 2 = 7$

      $\sqrt{24 + 1} + 2 = 7$

         $\sqrt{25} + 2 = 7$

           $5 + 2 = 7$

               $7 = 7$ A true statement

22. $\sqrt{3x + 1} + 6 = 2$

       $\sqrt{3x + 1} = -4$

    $(\sqrt{3x + 1})^2 = (-4)^2$ Square both sides

        $3x + 1 = 16$

           $3x = 15$

            $x = 5$

To check our solution when x = 5,

    $\sqrt{3(5) + 1} + 6 = 2$

      $\sqrt{15 + 1} + 6 = 2$

         $\sqrt{16} + 6 = 2$

           $4 + 6 = 2$

            $10 = 2$ A false statement

23. $$\sqrt{2x - 3} = x - 3$$
$$(\sqrt{2x - 3})^2 = (x - 3)^2 \qquad \text{Square both sides}$$
$$2x - 3 = x^2 - 6x + 9$$
$$0 = x^2 - 8x + 12$$
$$0 = (x - 6)(x - 2) \qquad \text{Factor}$$

$$0 = x - 6 \quad \text{or} \quad 0 = x - 2$$
$$6 = x \qquad\qquad 2 = x$$

To check our solution when x = 6,
$$\sqrt{2(6) - 3} = 6 - 3$$
$$\sqrt{12 - 3} = 3$$
$$\sqrt{9} = 3$$
$$3 = 3 \qquad \text{A true statement}$$

To check our solution when x = 2,
$$\sqrt{2(2) - 3} = 2 - 3$$
$$\sqrt{4 - 3} = -1$$
$$\sqrt{1} = -1$$
$$1 = -1 \qquad \text{A false statement}$$

24. Let x = a number
$$x - 4 = 3\sqrt{x}$$
$$(x - 4)^2 = (3\sqrt{x})^2 \qquad \text{Square both sides}$$
$$(x - 4)(x - 4) = (3\sqrt{x})(3\sqrt{x})$$
$$x^2 - 8x + 16 = 9x$$
$$x^2 - 17x + 16 = 0$$
$$(x - 16)(x - 1) = 1 \qquad \text{Factor}$$

$$x - 16 = 0 \quad \text{or} \quad x - 1 = 0$$
$$x = 16 \qquad\qquad x = 1$$

To check our solution when x = 16,
$$16 - 4 = 3\sqrt{16}$$
$$12 = 3(4)$$
$$12 = 12 \qquad \text{A true statement}$$

To check our solution when x = 1,
$$1 - 4 = 3\sqrt{1}$$
$$-3 = 3 \qquad \text{A false statement}$$

25. $y = \sqrt{x} - 2$

x	y
0	$\sqrt{0} - 2 = -2$
1	$\sqrt{1} - 3 = -1$
4	$\sqrt{4} - 2 = 0$
9	$\sqrt{9} - 2 = 1$
16	$\sqrt{16} - 2 = 2$

See the graph on page A-33 in the textbook.

Chapter 8

Section 8.1

1. $x^2 = 9$

$\sqrt{x^2} = \pm\sqrt{9}$ Take the square root of both sides

$x = \pm 3$

The two solutions are 3 and -3.

5. $y^2 = 8$

$\sqrt{y^2} = \pm\sqrt{8}$ Take the square root of both sides

$y = \pm\sqrt{8}$

$y = \pm\sqrt{4 \cdot 2}$

$y = \pm\sqrt{4}\sqrt{2}$

$y = \pm 2\sqrt{2}$

The solutions are $2\sqrt{2}$ and $-2\sqrt{2}$.

9. $3a^2 = 54$

$a^2 = 18$ Divide by 3

$\sqrt{a^2} = \pm\sqrt{18}$ Take the square root of both sides

$a = \pm\sqrt{18}$

$a = \pm\sqrt{9 \cdot 2}$

$a = \pm\sqrt{9}\sqrt{2}$

$a = \pm 3\sqrt{2}$

The solutions are $3\sqrt{2}$ and $-3\sqrt{2}$.

13. $(y + 1)^2 = 25$

$\sqrt{(y + 1)^2} = \pm\sqrt{25}$ Take the square root of both sides

$y + 1 = \pm 5$

$y = -1 \pm 5$

which we can write as

$y = -1 + 5$ or $y = -1 - 5$

$y = 4$ $\qquad\qquad$ $y = -6$

Our solutions are 4 and -6.

17. $(x + 1)^2 = 50$

$\sqrt{(x + 1)^2} = \pm\sqrt{50}$ Take the square root of both sides

$x + 1 = \pm\sqrt{50}$

$x = -1 \pm \sqrt{50}$

$x = -1 \pm \sqrt{25 \cdot 2}$

$x = -1 \pm \sqrt{25}\sqrt{2}$

$x = -1 \pm 5\sqrt{2}$

The two solutions are $-1 + 5\sqrt{2}$ and $-1 - 5\sqrt{2}$.

21. $(4a - 5)^2 = 26$

$\sqrt{(4a - 5)^2} = \pm\sqrt{36}$ Take the square root of both sides

$4a - 5 = \pm6$

$4a = 5 \pm 6$

$a = \dfrac{5 \pm 6}{4}$

Which we can write as

$a = \dfrac{5 + 6}{4}$ or $a = \dfrac{5 - 6}{4}$

$a = \dfrac{11}{4}$ $a = -\dfrac{1}{4}$

Our solutions are $\dfrac{11}{4}$ and $-\dfrac{1}{4}$.

25. $(6x + 2)^2 = 27$

$\sqrt{(6x + 2)^2} = \pm\sqrt{27}$ Take the square root of both sides

$6x + 2 = \pm\sqrt{9 \cdot 3}$

$6x + 2 = \pm\sqrt{9}\sqrt{3}$

$6x + 2 = \pm3\sqrt{3}$

$6x = -2 \pm 3\sqrt{3}$

$x = \dfrac{-2 \pm 3\sqrt{3}}{6}$

which we can write as

$x = \dfrac{-2 + 3\sqrt{3}}{6}$ or $x = \dfrac{-2 - 3\sqrt{3}}{6}$

Section 8.1 continued

29.
$$(3x + 6)^2 = 45$$
$$\sqrt{(3x + 6)^2} = \pm\sqrt{45}$$
$$3x + 6 = \pm\sqrt{9 \cdot 5}$$
$$3x + 6 = \pm\sqrt{9}\sqrt{5}$$
$$3x + 6 = \pm 3\sqrt{5}$$
$$3x = -6 \pm 3\sqrt{5}$$
$$x = \frac{-6 \pm 3\sqrt{5}}{3}$$

which we can write as

$$x = \frac{-6 + 3\sqrt{5}}{3} \quad \text{or} \quad x = \frac{-6 - 3\sqrt{5}}{3}$$

$$x = \frac{\cancel{3}(-2) + \sqrt{5})}{\cancel{3}} \qquad x = \frac{\cancel{3}(-2 - \sqrt{5})}{\cancel{3}}$$

$$x = -2 + \sqrt{5} \quad \text{or} \quad x = -2 - \sqrt{5}$$

33.
$$\left(x - \frac{2}{3}\right)^2 = \frac{25}{9}$$

$$x - \frac{2}{3} = \pm\frac{5}{3} \qquad \text{Square root property}$$

$$x = \frac{2}{3} \pm \frac{5}{3}$$

Which we can write as

$$x = \frac{2}{3} + \frac{5}{3} \quad \text{or} \quad x = \frac{2}{3} - \frac{5}{3}$$

$$x = \frac{7}{3} \qquad\qquad x = -\frac{3}{3}$$

$$x = -1$$

Our solutions are $\frac{7}{3}$ and -1.

37.
$$\left(a - \frac{4}{5}\right)^2 = \frac{12}{25}$$

$$a - \frac{4}{5} = \pm\frac{\sqrt{12}}{5} \qquad \text{Square root property}$$

$$a = \frac{4}{5} \pm \frac{2\sqrt{3}}{5} \qquad \sqrt{12} = \sqrt{4 \cdot 3} = 2\sqrt{3}$$

$$a = \frac{4 \pm 2\sqrt{3}}{5}$$

Our solutions are $\frac{4 + 2\sqrt{3}}{5}$, and $\frac{4 - 2\sqrt{3}}{5}$.

Section 8.1 continued

41. $x^2 - 2x + 1 = 9$
$(x - 1)^2 = 9$
$x - 1 = \pm 3$ Square root method
$x = 1 \pm 3$

Which we can write as

$x = 1 + 3$ or $x = 1 - 3$
$x = 4$ $x = -2$

Our solutions are -2 and 4.

45. When $x = -1 + 5\sqrt{2}$

the equation $(x + 1)^2 = 50$ becomes

$(-1 + 5\sqrt{2} + 1)^2 = 50$
$(5\sqrt{2})^2 = 50$
$(5\sqrt{2})(5\sqrt{2}) = 50$
$25(2) = 50$
$50 = 50$ A true statement

49. Given $A = 100(1 + r)^2$, solve for r

$A = 100(r + 1)^2$ Original formula

$\dfrac{A}{100} = (1 + r)^2$ Divide both sides by 100

$\dfrac{\sqrt{A}}{10} = (1 + r)$ Take the square root of each side

Since r represents interest rate, r will never be negative.

$\dfrac{\sqrt{A}}{10} = 1 + r$

$\dfrac{\sqrt{A}}{10} - 1 = r$ Add -1 to both sides

53. $(x - 5)^2 = (x - 5)(x - 5)$
$= x^2 + 2(x)(-5) + 25$ See page _____ in your textbook
$= x^2 - 10x + 25$

57. $x^2 + 4x + 4 = (x + 2)(x + 2)$
$= (x + 2)^2$

Notice that the first and last terms are perfect squares $(x)^2$ and $(2)^2$. Before going through the method for factoring trinomials by listing all possible factors, we can check to see if the square roots are factors.

61. $\sqrt[4]{16} = 2$ Because $(2)^4 = 16$

Section 8.2

1. Half of 6 is 3, the square of which is 9. If we add 9 to the end, we have
$$x^2 + 6x + 9 = (x + 3)^2$$

5. Half of -8 is -4, the square of which is 16. If we add 16 to the end, we have
$$y^2 - 8y + 16 = (y - 4)^2$$

9. Half of 16 is 8, the square of which is 64. If we add 64 to the end, we have
$$x^2 + 16x + 64 = (x + 8)^2$$

13. Half of -7 is $-\frac{7}{2}$, the square of which is $\frac{49}{4}$. If we add $\frac{49}{4}$ to the end we have
$$x^2 - 7x + \frac{49}{4} = \left(x - \frac{7}{2}\right)^2$$

17. Half of $-\frac{3}{2}$ is $-\frac{3}{4}$, the square of which is $\frac{9}{16}$. If we add $\frac{9}{16}$ to the end, we have
$$x^2 - \frac{3}{2}x + \frac{9}{16} = \left(x - \frac{3}{4}\right)^2$$

21.
$$x^2 - 6x = 16$$
$$x^2 - 6x + \mathbf{9} = 16 + \mathbf{9} \qquad \text{Half of 6 squared is 9}$$
$$(x - 3)^2 = 25$$
$$x - 3 = \pm 5 \qquad \text{Square root of both sides}$$
$$x = 3 \pm 5$$

$x = 3 + 5$ or $x = 3 - 5$
$x = 8 \qquad\qquad x = -2$

The solutions are 8 and -2.

25.
$$x^2 - 10x = 0$$
$$x^2 - 10x + \mathbf{25} = 0 + \mathbf{25} \qquad \text{Complete the square}$$
$$(x - 5)^2 = 25$$
$$x - 5 = \pm 5 \qquad \text{Square root of both sides}$$
$$x = 5 \pm 5 \qquad \text{Add 5 to both sides}$$

$x = 5 + 5$ or $x = 5 - 5$
$x = 10 \qquad\qquad x = 0$

The solutions are 10 and 0.

Section 8.2 continued

29.

$$x^2 + 4x - 3 = 0$$
$$x^2 + 4x = 3 \qquad \text{Add 3 to both sides}$$
$$x^2 + 4x + \mathbf{4} = 3 + \mathbf{4} \qquad \text{Complete the square}$$
$$(x + 2)^2 = \sqrt{7}$$
$$x + 2 = \pm\sqrt{7} \qquad \text{Square root of both sides}$$
$$x = -2 \pm \sqrt{7} \qquad \text{Add -2 to both sides}$$

$$x = -2 + \sqrt{7} \quad \text{or} \quad x = -2 - \sqrt{7}$$

The solutions written in a shorter form are $-2 \pm \sqrt{7}$.

33.

$$a^2 = 7a + 8$$
$$a^2 - 7a = 8 \qquad \text{Add -7a to both sides}$$
$$a^2 - 7a + \frac{49}{4} = 8 + \frac{49}{4} \qquad \text{Complete the square}$$
$$\left(a - \frac{7}{2}\right)^2 = \frac{81}{4} \qquad 8 + \frac{49}{4} = \frac{32}{4} + \frac{49}{4} = \frac{81}{4}$$
$$a - \frac{7}{2} = \pm\frac{9}{2} \qquad \text{Square root of both sides}$$
$$a = \frac{7}{2} \pm \frac{9}{2}$$

$$a = \frac{7}{2} + \frac{9}{2} \quad \text{or} \quad a = \frac{7}{2} - \frac{9}{2}$$
$$a = \frac{16}{2} \qquad\qquad a = -\frac{2}{2}$$
$$a = 8 \qquad\qquad a = -1$$

The solution are 8 and -1.

37.

$$2x^2 + 2x - 4 = 0$$
$$2x^2 + 2x = 4 \qquad \text{Add 4 to both sides}$$
$$x^2 + x = 2 \qquad \text{Divide by 2}$$
$$x^2 + x + \frac{1}{4} = 2 + \frac{1}{4} \qquad \text{Complete the square}$$
$$\left(x + \frac{1}{2}\right)^2 = \frac{9}{4} \qquad 2 + \frac{1}{4} = \frac{8}{4} + \frac{1}{4} = \frac{9}{4}$$
$$x + \frac{1}{2} = \pm\frac{3}{2} \qquad \text{Square root of both sides}$$
$$x = -\frac{1}{2} + \frac{3}{3} \qquad \text{Add } -\frac{1}{2} \text{ to both sides}$$

$$x = -\frac{1}{2} + \frac{3}{2} \quad \text{or} \quad x = -\frac{1}{2} - \frac{3}{2}$$
$$x = \frac{2}{2} \qquad\qquad x = -\frac{4}{2}$$
$$x = 1 \qquad\qquad x = -2$$

The solutions are 1 and -2.

41.

$$2x^2 - 2x = 1$$

$$x^2 - x = \frac{1}{2} \qquad \text{Divide by 2}$$

$$x^2 - x + \frac{1}{4} = \frac{1}{2} + \frac{1}{4} \qquad \text{Complete the square}$$

$$\left(x - \frac{1}{2}\right)^2 = \frac{3}{4} \qquad\qquad \frac{1}{2} + \frac{1}{4} = \frac{2}{4} + \frac{1}{4} = \frac{3}{4}$$

$$x - \frac{1}{2} = \pm\frac{\sqrt{3}}{2}$$

$$x = \frac{1}{2} \pm \frac{\sqrt{3}}{2} \qquad \text{Add } \frac{1}{2} \text{ to both sides}$$

$$x = \frac{1}{2} + \frac{\sqrt{3}}{2} \quad \text{or} \quad x = \frac{1}{2} - \frac{\sqrt{3}}{2}$$

$$x = \frac{1 + \sqrt{3}}{2} \qquad\qquad x = \frac{1 - \sqrt{3}}{2}$$

The solutions written in a shorter form are $\dfrac{1 \pm \sqrt{3}}{2}$.

45.

$$3y^2 - 9y = 2$$

$$y^2 - 3y = \frac{2}{3} \qquad \text{Divide by 3}$$

$$y^2 - 3y + \frac{9}{4} = \frac{2}{3} + \frac{9}{4} \qquad \text{Complete the square}$$

$$\left(y - \frac{3}{2}\right)^2 = \frac{35}{12} \qquad\qquad \frac{2}{3} + \frac{9}{4} = \frac{8}{12} + \frac{27}{12} = \frac{35}{12}$$

$$y - \frac{3}{2} = \pm\frac{\sqrt{35}}{2\sqrt{3}} \qquad\qquad \frac{\sqrt{35}}{2\sqrt{3}} \cdot \frac{\sqrt{3}}{\sqrt{3}} = \frac{\sqrt{105}}{6}$$

$$y = \frac{3}{2} \pm \frac{\sqrt{105}}{6}$$

$$y = \frac{3}{2} + \frac{\sqrt{105}}{6} \quad \text{or} \quad y = \frac{3}{2} - \frac{\sqrt{105}}{6}$$

$$y = \frac{9}{6} + \frac{\sqrt{105}}{6} \qquad\qquad y = \frac{9}{6} - \frac{\sqrt{105}}{6}$$

$$y = \frac{9 + \sqrt{105}}{6} \qquad\qquad y = \frac{9 - \sqrt{105}}{6}$$

The solutions written in shorter form are $\dfrac{9 \pm \sqrt{105}}{6}$.

Section 8.2 continued

49. When $x = -1 + \sqrt{2}$

the equation $4x^2 + 8x - 4 = 0$ becomes
$$4(-1 + \sqrt{2})^2 + 8(-1 + \sqrt{2}) - 4 = 0$$
$$4(1 - 2\sqrt{2} + 2) + 8(-1 + \sqrt{2}) - 4 = 0$$
$$12 - 8\sqrt{2} - 8 + 8\sqrt{2} - 4 = 0$$
$$0 = 0 \qquad \text{A true statement}$$

53. When $a = 2$,

the expression $2a$ becomes $2a = 2(2) = 4$

57. When $a = 2$, $b = 4$, and $c = -3$

the expression $\sqrt{b^2 - 4ac}$

becomes $\sqrt{(4)^2 - 4(2)(-3)} = \sqrt{16 + 24}$
$$= \sqrt{40}$$
$$= \sqrt{4 \cdot 10}$$
$$= 2\sqrt{10}$$

61. $\sqrt{20x^2y^3} = \sqrt{4x^2y^2 \cdot 5y}$
$$= 2xy\sqrt{5y}$$

Section 8.3

1. For the equation $x^2 + 3x + 2 = 0$, when $a = 1$, $b = 3$, and $c = 2$,
$$x = \frac{-b \pm \sqrt{b^2 - 4ac}}{2a}, \qquad \text{the quadratic formula becomes:}$$
$$= \frac{-3 \pm \sqrt{(3)^2 - 4(1)(2)}}{2(1)}$$
$$= \frac{-3 \pm \sqrt{9 - 8}}{2}$$
$$= \frac{-3 \pm 1}{2}$$

$x = \dfrac{-3 + 1}{2}$ or $x = \dfrac{-3 - 1}{2}$

$x = \dfrac{-2}{2} \qquad\qquad x = \dfrac{-4}{2}$

$x = -1 \qquad\qquad\quad x = -2$

The two solutions are -1 and 2.

Section 8.3 continued

5. For the equation $x^2 + 6x + 9 = 0$, when $a = 1$, $b = 6$, and $c = 9$,

$x = \dfrac{-b \pm \sqrt{b^2 - 4ac}}{2a}$, the quadratic formula, becomes:

$= \dfrac{-6 \pm \sqrt{(6)^2 - 4(1)(9)}}{2(1)}$

$= \dfrac{-6 \pm \sqrt{36 - 36}}{2}$

$= \dfrac{-6}{2}$

$= -3$

There is only one solution, -3.

9. For the equation $2x^2 + 5x + 3 = 0$, when $a = 2$, $b = 5$, and $c = 3$,

$x = \dfrac{-b \pm \sqrt{b^2 - 4ac}}{2a}$, the quadratic formula, becomes:

$= \dfrac{-5 \pm \sqrt{(5)^2 - 4(2)(3)}}{2(2)}$

$= \dfrac{-5 \pm \sqrt{25 - 24}}{4}$

$= \dfrac{-5 \pm 1}{4}$

$x = \dfrac{-5 + 1}{4}$ or $x = \dfrac{-5 - 1}{4}$

$x = \dfrac{-4}{4}$ $\qquad x = \dfrac{-6}{4}$

$x = -1$ $\qquad x = -\dfrac{3}{2}$

The two solutions are -1 and $-\dfrac{3}{2}$.

13. For the equation $x^2 - 2x + 1 = 0$, when $a = 1$, $b = -2$, and $c = 1$,

$x = \dfrac{-b \pm \sqrt{b^2 - 4ac}}{2a}$, the quadratic formula, becomes:

$= \dfrac{-(-2) \pm \sqrt{(-2)^2 - 4(1)(1)}}{2(1)}$

$= \dfrac{2 \pm \sqrt{4 - 4}}{2}$

$= \dfrac{2}{2}$

$= 1$

There is only one solution, 1.

17. For the equation $6x^2 - x - 2 = 0$, when $a = 6$, $b = -1$, and $c = -2$,

$$x = \frac{-b \pm \sqrt{b^2 - 4ac}}{2a},$$ the quadratic formula, becomes:

$$= \frac{-(-1) \pm \sqrt{(-1)^2 - 4(6)(-2)}}{2(6)}$$

$$= \frac{1 \pm \sqrt{1 + 48}}{12}$$

$$= \frac{1 \pm 7}{12}$$

$$x = \frac{1 + 7}{12} \quad \text{or} \quad x = \frac{1 - 7}{12}$$

$$x = \frac{8}{12} \qquad\qquad x = \frac{-6}{12}$$

$$x = \frac{2}{3} \qquad\qquad x = {}^-\frac{1}{2}$$

The two solutions are $\frac{2}{3}$ and $-\frac{1}{2}$.

21. $(2x - 3)(x + 2) = 1$
    $2x^2 + x - 6 = 1$ Multiply by FOIL
    $2x^2 + x - 7 = 0$ Standard form

When $a = 2$, $b = 1$, and $c = -7$,

$$x = \frac{-b \pm \sqrt{b^2 - 4ac}}{2a},$$ the quadratic formula, becomes:

$$= \frac{-1 \pm \sqrt{(1)^2 - 4(2)(-7)}}{2(2)}$$

$$= \frac{-1 \pm \sqrt{1 + 56}}{4}$$

$$= \frac{-1 \pm \sqrt{57}}{4}$$

The two solutions are $\frac{-1 \pm \sqrt{57}}{4}$.

25.
$$2x^2 = -6x + 7$$
$$2x^2 + 6x - 7 = 0 \qquad \text{Standard form}$$

When a = 2, b = 6, and c = -7:

$$x = \frac{-b \pm \sqrt{b^2 - 4ac}}{2a}, \qquad \text{the quadratic formula, becomes:}$$

$$= \frac{-6 \pm \sqrt{(6)^2 - 4(2)(-7)}}{2(2)}$$

$$= \frac{-6 \pm \sqrt{36 + 56}}{4}$$

$$= \frac{-6 \pm \sqrt{96}}{4}$$

$$= \frac{-6 \pm \sqrt{4 \cdot 23}}{4}$$

$$= \frac{-6 \pm 2\sqrt{23}}{4}$$

$$= \frac{2(-3) \pm \sqrt{23}}{4}$$

$$= \frac{-3 \pm \sqrt{23}}{2}$$

The two solutions are $\dfrac{-3 \pm \sqrt{23}}{2}$.

29.
$$2x^2 - 5 = 2x$$
$$2x^2 - 2x - 5 = 0 \qquad \text{Standard form}$$

When a = 2, b = -2, and c = -5:

$$x = \frac{-b \pm \sqrt{b^2 - 4ac}}{2a}, \qquad \text{the quadratic formula, becomes:}$$

$$x = \frac{-(-2) \pm \sqrt{(-2)^2 - 4(2)(-5)}}{2(2)}$$

$$= \frac{2 \pm \sqrt{4 + 40}}{4}$$

$$= \frac{2 \pm \sqrt{4 \cdot 11}}{4}$$

$$= \frac{2 \pm 2\sqrt{11}}{4}$$

$$= \frac{2(1 \pm \sqrt{11})}{4}$$

$$= \frac{1 \pm \sqrt{11}}{2}$$

The two solutions are $\dfrac{1 \pm \sqrt{11}}{2}$.

33. For the equation $3x^2 - 4x = 0$, when $a = 3$, $b = -4$, and $c = 0$,

$$x = \frac{-b \pm \sqrt{b^2 - 4ac}}{2a} \quad \text{the quadratic formula, becomes:}$$

$$= \frac{-(-4) \pm \sqrt{(-4)^2 - 4(3)(0)}}{2(3)}$$

$$= \frac{4 \pm \sqrt{16}}{6}$$

$$= \frac{4 \pm 4}{6}$$

$$x = \frac{4 + 4}{6} \quad \text{or} \quad x = \frac{4 - 4}{6}$$

$$x = \frac{8}{6} \qquad\qquad x = \frac{0}{6}$$

$$x = \frac{4}{3} \qquad\qquad x = 0$$

37. $(2\sqrt{3})(3\sqrt{5}) = (2 \cdot 3)(\sqrt{3} \cdot \sqrt{5})$
$$= 6\sqrt{15}$$

41. $(\sqrt{7} - \sqrt{2})(\sqrt{7} + \sqrt{2}) = (\sqrt{7})^2 - (\sqrt{2})^2 \qquad (a - b)(a + b) = a^2 + b^2$
$$= 7 - 2$$
$$= 5$$

Section 8.4

1. $(3 - 2i) + 3i = 3 + (-2i + 3i)$ Associative property
$$= 3 + i \qquad\qquad\qquad\quad \text{Combine similar terms}$$

5. $(11 + 9i) - 9i = 11 + (9i - 9i)$ Associative property
$$= 11$$

9. $(5 + 7i) - (6 + 8i) = 5 + 7i - 6 - 8i$ Distributive property
$$= (5 - 6) + (7i - 8i) \qquad \text{Commutative and Associative properties}$$
$$= -1 - i$$

13. $(6 + i) - 4i - (2 - i) = 6 + i - 4i - 2 + i$ Distributive property
$$= (6 - 2) + (i - 4i + i) \qquad \text{Commutative and Associative properties}$$
$$= 4 - 2i$$

17. $(2 + 3i) - (6 - 2i) + (3 - i) = 2 + 3i - 6 + 2i + 3 - i$ Distributive property

$$= (2 - 6 + 3) + (3i + 2i - i)$$
$$= -1 + 4i$$

21. $2i(8 - 7i) = 2i(8) - 2i(7i)$ Distributive property
$$= 16i - 14i^2$$
$$= 16i - 14(-1) \qquad i^2 = 1$$
$$= 14 + 16i$$

25. $(2 + i)(3 - 5i) = 2 \quad 3 + 2(-5i) + i(3) + i(-5i)$ FOIL method
$$= 6 - 10i + 3i - 5i^2$$
$$= 6 - 7i - 5(-1) \qquad\qquad i^2 = -1$$
$$= 11 - 7i$$

29. $(2 + i)(2 - i) = (2)^2 - (i)^2$
$$= 4 - (-1)$$
$$= 5$$

33. $\dfrac{-3i}{2 + 3i} \left(\dfrac{\mathbf{2 - 3i}}{\mathbf{2 - 3i}} \right) = \dfrac{-6i + 9i^2}{4 - 9i^2}$

$$= \dfrac{-6i - 9}{4 + 9}$$

$$= \dfrac{-6i - 9}{13}$$

$$= \dfrac{-9 - 6i}{13}$$

37. $\dfrac{2 + i}{2 - i} \dfrac{\mathbf{2 + i}}{\mathbf{2 + i}} = \dfrac{4 + 2(2i) + i^2}{4 - i^2}$

$$= \dfrac{4 + 4i - 1}{4 - (-1)}$$

$$= \dfrac{3 + 4i}{5}$$

41. $(x + 3i)(x - 3i) = x(x) + x(-3i) + 3i(x) + 3i(3i)$

                          F   **O**   **I**   **L**

$$= x^2 - 3xi + 3xi - 9i^2$$
$$= x^2 - 9(-1)$$
$$= x^2 + 9$$

45. $\qquad\qquad (x - 3)^2 = 25$
$$x - 3 = \pm 5 \qquad \text{Square root of each side}$$
$$x = 3 \pm 5 \qquad \text{Add 3 to both sides}$$

$x = 3 + 5$ or $x = 3 - 5$
$x = 8$ $x = -2$

The two solutions are 8 and -2.

49. $(x + 3)^2 = 12$

$x + 3 = \pm\sqrt{12}$ Square root of each side

$x + 3 = \pm 2\sqrt{3}$ $\sqrt{12} = \sqrt{4 \cdot 3} = 2\sqrt{3}$

$x = -3 \pm 2\sqrt{3}$

53. $\sqrt{\dfrac{8x^2y^3}{3}} = \dfrac{\sqrt{8x^2y^3}}{\sqrt{3}}$

$= \dfrac{\sqrt{4x^2y^2 \cdot 2y}}{3}$

$= \dfrac{2xy\sqrt{2y}}{3}$

$= \dfrac{2xy\sqrt{2y}}{3} \cdot \dfrac{\sqrt{3}}{\sqrt{3}}$

$= \dfrac{2xy\sqrt{6y}}{3}$

Section 8.5

1. $\sqrt{-16} = \sqrt{16(-1)} = \sqrt{16}\sqrt{-1} = 4i$

5. $\sqrt{-6} = \sqrt{6(-1)} = \sqrt{6}\sqrt{-1} = i\sqrt{6}$

9. $\sqrt{-32} = \sqrt{32(-1)} = \sqrt{32}\sqrt{-1} = 4i\sqrt{2}$

13. $\sqrt{-8} = \sqrt{8(-1)} = \sqrt{8}\sqrt{-1} = 2i\sqrt{2}$

17. $x^2 = 2x - 2$

$x^2 - 2x + 2 = 0$ Standard form

When $a = 1$, $b = -2$, and $c = 2$,

$x = \dfrac{-b \pm \sqrt{b^2 - 4ac}}{2a}$,   the quadratic formula, becomes:

$= \dfrac{-(-2) \pm \sqrt{(-2)^2 - 4(1)(2)}}{2(1)}$

$= \dfrac{2 \pm \sqrt{4 - 8}}{2}$

$= \dfrac{2 \pm \sqrt{-4}}{2}$

$= \dfrac{2 \pm 2i}{2}$

$= \dfrac{2(1 \pm i)}{2}$

$= 1 \pm i$

The two solutions are $1 + i$ and $1 - i$.

21. 　　　$2x^2 + 5x = 12$
　　　$2x^2 + 5x - 12 = 0$　　　Standard form

When $a = 2$, $b = 5$, and $c = -12$,

$$x = \frac{-b \pm \sqrt{b^2 - 4ac}}{2a},$$ 　　the quadratic formula, becomes:

$$= \frac{-5 \pm \sqrt{(5)^2 - 4(2)(-12)}}{2(2)}$$

$$= \frac{-5 \pm \sqrt{25 + 96}}{4}$$

$$= \frac{-5 \pm \sqrt{121}}{4}$$

$$= \frac{-5 \pm 11}{4}$$

$x = \dfrac{-5 + 11}{4}$　or　$x = \dfrac{-5 - 11}{4}$

$x = \dfrac{6}{4}$　　　　　　　$x = \dfrac{-16}{4}$

$x = \dfrac{3}{2}$　　　　　　　$x = -4$

The two solutions are $\dfrac{3}{2}$ and -4.

25. 　$(x + \frac{1}{2})^2 = -\dfrac{9}{4}$

$x + \dfrac{1}{2} = \pm \sqrt{-\dfrac{9}{4}}$　　Square root of both sides

$x + \dfrac{1}{2} = \pm \dfrac{3}{2}i$　　₈　$\sqrt{-\dfrac{9}{4}} = \sqrt{\dfrac{9}{4}}\sqrt{-1} = \dfrac{3}{2}i$

$x = -\dfrac{1}{2} \pm \dfrac{3i}{2}$

$x = \dfrac{-1 \pm 3i}{2}$

The two solutions are $\dfrac{-1 + 3i}{2}$ and $\dfrac{-1 - 3i}{2}$.

29. $x^2 + x + 1 = 0$

When $a = 1$, $b = 1$, and $c = 1$,

$$x = \frac{-b \pm \sqrt{b^2 - 4ac}}{2a}, \qquad \text{the quadratic formula, becomes:}$$

$$= \frac{-1 \pm \sqrt{(1)^2 - 4(1)(-1)}}{2(1)}$$

$$= \frac{-1 \pm \sqrt{1 - 4}}{2}$$

$$= \frac{-1 \pm \sqrt{-3}}{2}$$

$$= \frac{-1 \pm i\sqrt{3}}{2}$$

The two solutions are $\frac{-1 + i\sqrt{3}}{2}$ and $\frac{-1 - i\sqrt{3}}{2}$.

33. $\frac{1}{2}x^2 + \frac{1}{3}x + \frac{1}{6} = 0$

$3x^2 + 2x + 1 = 0 \qquad$ Multiply both sides by LCD which is 6.

When $a = 3$, $b = 2$, and $c = 1$,

$$x = \frac{-b \pm \sqrt{b^2 - 4ac}}{2a}, \qquad \text{the quadratic formula, becomes:}$$

$$= \frac{-2 \pm \sqrt{2^2 - 4(3)(1)}}{2(3)}$$

$$= \frac{-2 \pm \sqrt{4 - 12}}{6}$$

$$= \frac{-2 \pm \sqrt{-8}}{6}$$

$$= \frac{-2 \pm \sqrt{4(2)(-1)}}{6}$$

$$= \frac{-2 \pm 2i\sqrt{2}}{6}$$

$$= \frac{2(-1 \pm i\sqrt{2})}{6}$$

$$= \frac{-1 \pm i\sqrt{2}}{3}$$

The two solutions are $\frac{-1 + i\sqrt{2}}{3}$ and $\frac{-1 - i\sqrt{2}}{3}$.

37. $(x + 2)(x - 3) = 5$
$\qquad x^2 - x - 6 = 5 \qquad$ FOIL method
$\qquad x^2 - x - 11 = 0 \qquad$ Add -5 to both sides, standard form

When $a = 1$, $b = -1$, and $c = -11$,

$\qquad x = \dfrac{-b \pm \sqrt{b^2 - 4ac}}{2a}$, $\qquad$ the quadratic formula, becomes:

$$= \dfrac{-(-1) \pm \sqrt{(-1)^2 - 4(1)(-11)}}{2(1)}$$

$$= \dfrac{1 \pm \sqrt{1 + 44}}{2}$$

$$= \dfrac{1 \pm \sqrt{45}}{2}$$

$$= \dfrac{1 \pm 3\sqrt{5}}{2}$$

The two solutions are $\dfrac{1 + 3\sqrt{5}}{2}$ and $\dfrac{1 - 3\sqrt{5}}{2}$.

41. $(2x - 2)(x - 3) = 9$
$\qquad 2x^2 - 8x + 6 = 9 \qquad$ FOIL method
$\qquad 2x^2 - 8x - 3 = 0 \qquad$ Add -9 to both sides, standard form

When $a = 2$, $b = -8$, and $c = -3$,

$\qquad x = \dfrac{-b \pm \sqrt{b^2 - 4ac}}{2a}$, $\qquad$ the quadratic formula, becomes:

$$= \dfrac{-(-8) \pm \sqrt{(-8)^2 - 4(2)(-3)}}{2(2)}$$

$$= \dfrac{8 \pm \sqrt{64 + 24}}{4}$$

$$= \dfrac{8 \pm \sqrt{88}}{4}$$

$$= \dfrac{8 \pm 2\sqrt{22}}{4}$$

$$= \dfrac{2(4 \pm \sqrt{22})}{4}$$

$$= \dfrac{4 \pm \sqrt{22}}{2}$$

45. $3 - 7i$ because both solutions would be $3 \pm 7i$.

Section 8.5 continued

49. Graph $2x + 4y = 8$

x-intercept
when $\qquad y = 0$
the equation $\quad 2x + 4y = 8$
becomes $\qquad 2x + 4(0) = 8$
$\qquad\qquad\qquad 2x = 8$
$\qquad\qquad\qquad x = 4$

y-intercept
when $\qquad x = 0$
the equation $\quad 2x + 4y = 8$
becomes $\qquad 2(0) + 4y = 8$
$\qquad\qquad\qquad 4y = 8$
$\qquad\qquad\qquad y = 2$

The x-intercept is at $(4,0)$.   The y-intercept is at $(0,2)$.

See the graph in your textbook page A-34.

53.

x	$y = x^2 + 2$	y
-2	$y = (-2)^2 + 2$	6
-1	$y = (-1)^2 + 2$	3
0	$y = 0^2 + 2$	2
1	$y = 1^2 + 2$	3
2	$y = 2^2 + 2$	6

See the graph on page A-34 in the textbook.

57.
$$\sqrt{-24} - \sqrt{54} - \sqrt{150} = \sqrt{4 \cdot 6} - \sqrt{9 \cdot 6} - \sqrt{25 \cdot 6}$$
$$= 2\sqrt{6} - 3\sqrt{6} - 5\sqrt{6}$$
$$= -6\sqrt{6}$$

Section 8.6

1.

x	$y = x^2 - 4$	y
-3	$y = (-3)^2 - 4$	5
-2	$y = (-2)^2 - 4$	0
-1	$y = (-1)^2 - 4$	-3
0	$y = (0)^2 - 4$	-4
1	$y = (1)^2 - 4$	-3
2	$y = (2)^2 - 4$	0

See the graph on page A-35 in the textbook.

5.

x	$y = (x + 2)^2$	y
-4	$y = (-4 + 2)^2$	4
-2	$y = (-2 + 2)^2$	0
0	$y = (0 + 2)^2$	4

See the graph on page A-35 in the textbook.

9.

x	$y = (x - 5)^2$	y
3	$y = (3 - 5)^2$	4
4	$y = (4 - 5)^2$	1
5	$y = (5 - 5)^2$	0
6	$y = (6 - 5)^2$	1

See the graph on page A-35 in the textbook.

13.

x	$y = (x + 2)^2 - 3$	y
-4	$y = (-4 + 2)^2 - 3$	1
-3	$y = (-3 + 2)^2 - 3$	-2
-2	$y = (-2 + 2)^2 - 3$	-3
-1	$y = (-1 + 2)^2 - 3$	-2
0	$y = (0 + 2)^2 - 3$	1

See the graph on page A-35 in the textbook.

17.
$$y = x^2 + 6x + 5$$
$$y = (x^2 + 6x + \mathbf{9}) + 5 - \mathbf{9}$$
$$y = (x + 3)^2 - \mathbf{4}$$

See the graph on page A-35 in the textbook.

21.

x	$y = 4 - x^2$	y
-3	$y = 4 - (-3)^2$	-5
-2	$y = 4 - (-2)^2$	0
-1	$y = 4 - (-1)^2$	3
0	$y = 4 - (0)^2$	4
1	$y = 4 - (1)^2$	3
2	$y = 4 - (2)^2$	0
3	$y = 4 - (3)^2$	-5

See the graph on page A-36 in the textbook.

Section 8.6 continued

25. To graph the line $y = x + 2$:

x-intercept

when	$y = 0$
the equation	$y = x + 2$
becomes	$0 = x + 2$
	$-2 = x$

The x-intercept is at $(-2, 0)$.

y-intercept

when	$x = 0$
the equation	$y = x + 2$
becomes	$y = 0 + 2$
	$y = 2$

The y-intercept is at $(0, 2)$.

To graph the parabola $y = x^2$:

x	$y = x^2$	y
-2	$y = (-2)^2$	4
-1	$y = (-1)^2$	1
0	$y = (0)^2$	0
1	$y = (1)^2$	1
2	$y = (2)^2$	4

See the graph on page A-36 in the textbook.

29. $\sqrt{x + 5} = 4$

$\quad x + 5 = 16 \qquad$ Square both sides

$\qquad\quad x = 11$

Check:

$$\sqrt{11 + 5} = 4$$

$$\sqrt{16} = 4$$

$$4 = 4 \qquad \text{A true statement}$$

33. $\sqrt{3a + 2} - 3 = 5$

$\sqrt{3a + 2} = 8$	Add 3 to both sides
$3a + 2 = 64$	Square both sides
$3a = 62$	Add -2 to each side
$a = \dfrac{62}{3}$	Divide 3 to each side

Chapter 8 Review

1. $a^2 = 32$

$a = \pm\sqrt{32}$

$a = \pm\sqrt{16 \cdot 2}$

$a = \pm 4\sqrt{2}$

The solutions are $-4\sqrt{2}$ and $4\sqrt{2}$.

5.

$(x + 3)^2 = 36$

$x + 3 = \pm 6$

$x = -3 \pm 6$

$x = -3 + 6$ or $x = -3 - 6$

$x = 3 \qquad\qquad x = -9$

The solutions are -9 and 3.

9. $(2x + 5)^2 = 32$

$2x + 5 = \pm\sqrt{32}$

$2x + 5 = \pm 4\sqrt{2}$ $\qquad\qquad \sqrt{32} = \sqrt{16 \cdot 2} = 4\sqrt{2}$

$2x = -5 \pm 4\sqrt{2}$

$x = \dfrac{-5 \pm 4\sqrt{2}}{2}$

The solutions are $\dfrac{-5 + 4\sqrt{2}}{2}$ and $\dfrac{-5 - 4\sqrt{2}}{2}$.

13. $x^2 + 8x = 4$

$x^2 + 8x + \dfrac{64}{4} = 4 + \dfrac{64}{4}$ $\qquad \left(\dfrac{8}{2}\right)^2 = \dfrac{64}{4}$

$(x + 4)^2 = 4 + 16$ $\qquad\qquad \dfrac{64}{4} = 16$

$x + 4 = \pm\sqrt{20}$ $\qquad\qquad$ Square root property

$x + 4 = \pm 2\sqrt{5}$ $\qquad\qquad \sqrt{20} = \sqrt{4 \cdot 5} = 2\sqrt{5}$

$x = -4 \pm 2\sqrt{5}$

The solutions are $-4 + 2\sqrt{5}$ and $-4 - 2\sqrt{5}$.

17.
$$a^2 = 9a + 3$$
$$a^2 - 9a = 3 \qquad \text{Add } -9a \text{ to both sides}$$
$$a^2 - 9a + \frac{81}{4} = 3 + \frac{81}{4} \qquad \left(\frac{9}{2}\right)^2 = \frac{81}{4}$$
$$\left(a - \frac{9}{2}\right)^2 = \frac{93}{4} \qquad 3 + \frac{81}{4} = \frac{12}{4} + \frac{81}{4} = \frac{93}{4}$$
$$a - \frac{9}{2} = \pm\sqrt{\frac{93}{4}}$$
$$a - \frac{9}{2} = \frac{\pm\sqrt{93}}{2}$$
$$a = \frac{9}{2} \pm \frac{\sqrt{93}}{2}$$
$$a = \frac{9 \pm \sqrt{93}}{2}$$

The solutions are $\dfrac{9 + \sqrt{93}}{2}$ and $\dfrac{9 - \sqrt{93}}{2}$.

21. $x^2 + 7x + 12 = 0$

Let $a = 1$, $b = 7$, $c = 12$, then

$$x = \frac{-b \pm \sqrt{b^2 - 4ac}}{2a}$$
$$= \frac{-7 \pm \sqrt{7^2 - 4(1)(12)}}{2(1)}$$
$$= \frac{-7 \pm \sqrt{49 - 48}}{2}$$
$$= \frac{-7 \pm 1}{2}$$

$$x = \frac{-7 + 1}{2} \quad \text{or} \quad x = \frac{-7 - 1}{2}$$
$$= \frac{-6}{2} \qquad\qquad = \frac{-8}{2}$$
$$= -3 \qquad\qquad\quad = -4$$

The two solutions are -4 and -3.

25.
$$2x^2 = -8x + 5$$
$$2x^2 + 8x - 5 = 0 \qquad \text{Standard form}$$

Let a = 2, b = 8, c = -5, then

$$x = \frac{-b \pm \sqrt{b^2 - 4ac}}{2a}$$

$$= \frac{-8 \pm \sqrt{8^2 - 4(2)(-5)}}{2(2)}$$

$$= \frac{-8 \pm \sqrt{64 + 40}}{4}$$

$$= \frac{-8 \pm \sqrt{104}}{4} \qquad\qquad \sqrt{104} = \sqrt{4 \cdot 26} = 2\sqrt{26}$$

$$= \frac{-8 \pm 2\sqrt{26}}{4}$$

$$= \frac{2(-4 \pm \sqrt{26})}{4}$$

$$= \frac{-4 \pm \sqrt{26}}{2}$$

The two solutions are $\frac{-4 + \sqrt{26}}{2}$ and $\frac{-4 - \sqrt{26}}{2}$.

29.
$$(2x + 1)(2x - 3) = -6$$
$$4x^2 - 4x - 3 = -6 \qquad \text{Foil Method}$$
$$4x^2 - 4x + 3 = 0 \qquad \text{Standard form}$$

Let a = 4, b = -4 and c = 3, then

$$x = \frac{-b \pm \sqrt{b^2 - 4ac}}{2a}$$

$$= \frac{-(-4) \pm \sqrt{(-4)^2 - 4(4)(3)}}{2(4)}$$

$$= \frac{4 \pm \sqrt{16 - 48}}{8}$$

$$= \frac{4 \pm \sqrt{-32}}{8}$$

$$= \frac{4 \pm 4i\sqrt{2}}{8}$$

$$= \frac{4(1 \pm i\sqrt{2})}{8}$$

$$= \frac{1 \pm i\sqrt{2}}{2}$$

The solutions are $\frac{1 + i\sqrt{2}}{2}$ and $\frac{1 - i\sqrt{2}}{2}$.

Chapter 8 Review continued

33. $(5 + 6i) + (5 - i) = (5 + 5) + (6i - i)$
$$= 10 + 5i$$

37. $(3 + i) - 5i - (4 - i) = 3 + i - 5i - 4 + i$
$$= (3 - 4) + (i - 5i + i)$$
$$= -1 - 3i$$

41. $2(3 - i) = 2(3) - 2(i)$
$$= 6 - 2i$$

45. $(2 - i)(3 + i) = 6 + 2i - 3i - i^2$
$$\qquad\qquad\quad \text{F}\quad\text{O}\quad\text{I}\quad\text{L}$$
$$= 6 - i + 1 \qquad\qquad -i^2 = -(-1) = 1$$
$$= 7 - i$$

49. $(4 + i)(4 - i) = 16 - i^2 \qquad$ Difference of two perfect squares
$$= 16 + 1 \qquad -i^2 = -(-1) = 1$$
$$= 17$$

53. $\dfrac{5}{2 + 5i} = \left(\dfrac{5}{2 + 5i}\right)\left(\dfrac{2 - 5i}{2 - 5i}\right)$

$$= \dfrac{10 - 25i}{4 - 25i^2} \qquad\qquad i^2 = -1$$

$$= \dfrac{10 - 25i}{4 + 25}$$

$$= \dfrac{10 - 25i}{29}$$

57. $\dfrac{3 + i}{3 - i} = \left(\dfrac{3 + i}{3 - i}\right)\left(\dfrac{3 + i}{3 + i}\right)$

$$= \dfrac{9 + 3i + 3i + i^2}{9 + i^2} \qquad i^2 = -1$$

$$= \dfrac{8 + 6i}{10}$$

$$= \dfrac{2(4 + 3i)}{10}$$

$$= \dfrac{4 + 3i}{5}$$

61. $\sqrt{-36} = \sqrt{36(-1)} = \sqrt{36}\,\sqrt{-1} = 6i$

65. $\sqrt{-40} = \sqrt{4(-1)(10)} = \sqrt{4}\,\sqrt{-1}\,\sqrt{10} = 2i\sqrt{10}$

Chapter 8 Review continued

69. $y = x^2 + 2$

x	y
-2	$(-2)^2 + 2 = 6$
-1	$(-1)^2 + 2 = 3$
0	$0^2 + 2 \quad = 2$
1	$1^2 + 2 \quad = 3$
2	$2^2 + 2 \quad = 6$

See the graph on page A-37 in the textbook.

73. $y = (x + 3)^2 - 2$

x	y
-5	$(-5 + 3)^2 - 2 = 2$
-4	$(-4 + 3)^2 - 2 = -1$
-3	$(-3 + 3)^2 - 2 = -2$
-2	$(-2 + 3)^2 - 2 = -1$
-1	$(-1 + 3)^2 - 2 = 2$

See the graph on page A-37 in the textbook.

Chapter 8 Test

1. For the equation $x^2 - 7x - 8 = 0$, $a = 1$, $b = -7$ and $c = -8$:

$$x = \frac{-b \pm \sqrt{b^2 - 4ac}}{2a}$$

$$= \frac{-(-7) \pm \sqrt{(-7)^2 - 4(1)(-8)}}{2(1)}$$

$$= \frac{7 \pm \sqrt{49 + 32}}{2}$$

$$= \frac{7 \pm 9}{2}$$

$$x = \frac{7 + 9}{2} \quad \text{or} \quad x = \frac{7 - 9}{2}$$

$$x = \frac{16}{2} \qquad\qquad x = \frac{-2}{2}$$

$$x = 8 \qquad\qquad x = -1$$

The two solutions are 8 and -1.

2.
$$(x - 3)^2 = 12$$
$$x^2 - 6x + 9 = 12 \qquad \text{FOIL Method}$$
$$x^2 - 6x - 3 = 0 \qquad \text{Add -12 to both sides}$$

$a = 1$, $b = -6$ and $c = -3$

$$x = \frac{-b \pm \sqrt{b^2 - 4ac}}{2a}$$

$$= \frac{-(-6) \pm \sqrt{(-6)^2 - 4(1)(-3)}}{2(1)}$$

$$= \frac{6 \pm \sqrt{36 + 12}}{2}$$

$$= \frac{6 \pm \sqrt{48}}{2}$$

$$= \frac{6 \pm 4\sqrt{3}}{2}$$

$$= \frac{\overset{1}{2}(3 \pm 2\sqrt{3})}{\underset{1}{2}}$$

$$= 3 \pm 2\sqrt{3}$$

Chapter 8 Test continued

3. $\left(x - \dfrac{5}{2}\right)^2 = -\dfrac{75}{4}$

$x - \dfrac{5}{2} = \pm\sqrt{-\dfrac{75}{4}}$

$x - \dfrac{5}{2} = \pm\dfrac{5i\sqrt{3}}{2}$ $\sqrt{-\dfrac{75}{4}} = \dfrac{\sqrt{25(-1)(3)}}{\sqrt{4}} = \dfrac{5i\sqrt{3}}{2}$

$x = \dfrac{5}{2} \pm \dfrac{5i\sqrt{3}}{2}$

$x = \dfrac{5 \pm 5i\sqrt{3}}{2}$

The solutions are $\dfrac{5 + 5i\sqrt{3}}{2}$ and $\dfrac{5 - 5i\sqrt{3}}{2}$.

4. $\dfrac{1}{3}x^2 = \dfrac{1}{2}x - \dfrac{5}{6}$

$6\left(\dfrac{1}{3}x^2\right) = 6\left(\dfrac{1}{2}x\right) - 6\left(\dfrac{5}{6}\right)$ LCD = 6

$2x^2 = 3x - 5$

$2x^2 - 3x + 5 = 0$ Standard form

$a = 2$, $b = -3$ and $c = 5$

$x = \dfrac{-b \pm \sqrt{b^2 - 4ac}}{2a}$

$= \dfrac{-(-3) \pm \sqrt{(-3)^2 - 4(2)(5)}}{2(2)}$

$= \dfrac{3 \pm \sqrt{9 - 40}}{4}$

$= \dfrac{3 \pm \sqrt{-31}}{4}$

$= \dfrac{3 \pm i\sqrt{31}}{4}$

5.
$$3x^2 = -2x + 1$$
$$3x^2 + 2x - 1 = 0$$

$a = 3$, $b = 2$ and $c = -1$

$$x = \frac{-b \pm \sqrt{b^2 - 4ac}}{2a}$$

$$= \frac{-(2) \pm \sqrt{(2)^2 - 4(3)(-1)}}{2(3)}$$

$$= \frac{-2 \pm \sqrt{4 + 12}}{6}$$

$$= \frac{-2 \pm 4}{6}$$

$$= \frac{\overset{1}{\cancel{2}}(-1 \pm 2)}{\underset{3}{\cancel{6}}}$$

$$x = \frac{-1 + 2}{3} \quad \text{or} \quad x = \frac{-1 - 2}{3}$$

$$x = \frac{1}{3} \qquad\qquad x = \frac{-3}{3}$$

$$\qquad\qquad\qquad x = -1$$

The two solutions are $\frac{1}{3}$ and -1.

6.
$$(x + 2)(x - 1) = 6$$
$$x^2 + x - 2 = 6 \qquad \text{FOIL Method}$$
$$x^2 + x - 8 = 0$$

$a = 1$, $b = 1$ and $c = -8$

$$x = \frac{-b \pm \sqrt{b^2 - 4ac}}{2a}$$

$$= \frac{-1 \pm \sqrt{(1)^2 - 4(1)(-8)}}{2(1)}$$

$$= \frac{-1 \pm \sqrt{1 + 32}}{2}$$

$$= \frac{-1 \pm \sqrt{33}}{2}$$

The solutions are $\frac{-1 + \sqrt{33}}{2}$ and $\frac{-1 - \sqrt{33}}{2}$.

7. $9x^2 + 12x + 4 = 0$

   $a = 9$, $b = 12$ and $c = 4$

   $$x = \frac{-b \pm \sqrt{b^2 - 4ac}}{2a}$$

   $$= \frac{-12 \pm \sqrt{(12)^2 - 4(9)(4)}}{2(9)}$$

   $$= \frac{-12 \pm \sqrt{144 - 144}}{18}$$

   $$= \frac{-12}{18}$$

   $$= \frac{-2}{3}$$

8. $x^2 - 6x - 6 = 0$

   $x^2 - 6x = 6$ Add 6 to both sides

   $x^2 - 6x + \mathbf{9} = 6 + \mathbf{9}$ Complete the square

   $(x - 3)^2 = 15$

   $x - 3 = \pm\sqrt{15}$

   $x = 3 \pm \sqrt{15}$

9. $\sqrt{-9} = \sqrt{9(-1)} = \sqrt{9}\,\sqrt{-1} = 3i$

10. $\sqrt{-121} = \sqrt{121(-1)} = \sqrt{121}\,\sqrt{-1} = 11i$

11. $\sqrt{-72} = \sqrt{72(-1)} = \sqrt{72}\,\sqrt{-1} = 6i\sqrt{2}$

12. $\sqrt{-18} = \sqrt{18(-1)} = \sqrt{18}\,\sqrt{-1} = 3i\sqrt{2}$

13. $(3i + 1) + (2 + 5i) = (1 + 2) + (3i + 5i)$

   $\qquad\qquad\qquad\qquad = 3 + 8i$

14. $(6 - 2i) - (7 - 4i) = 6 - 2i - 7 + 4i$

   $\qquad\qquad\qquad\qquad = (6 - 7) + (-2i + 4i)$

   $\qquad\qquad\qquad\qquad = -1 + 2i$

15. $(2 + i)(2 - i) = (2)^2 - (i)^2$

   $\qquad\qquad\qquad = 4 - (-1)$

   $\qquad\qquad\qquad = 5$

   Remember: $(a + bi)(a - bi) = (a)^2 - (bi)^2$

16. $(3 + 2i)(1 + i) = 3(1) + 3(i) + 2i(1) + 2i(i)$

   $\qquad\qquad\qquad\qquad$ F $\qquad$ O $\qquad$ I $\qquad$ L

   $\qquad\qquad = 3 + 3i + 2i + 2i^2$

   $\qquad\qquad = 3 + 5i + 2(-1)$

   $\qquad\qquad = 1 + 5i$

Chapter 8 Test continued

17.
$$\frac{i}{3 - i} = \frac{i}{3 - i}\left(\frac{3 + i}{3 + i}\right)$$

$$= \frac{3i + i^2}{9 - i^2}$$

$$= \frac{3i - 1}{9 - (-1)}$$

$$= \frac{-1 + 3i}{10}$$

18.
$$\frac{2 + i}{2 - i} = \frac{2 + i}{2 - i}\left(\frac{2 + i}{2 + i}\right)$$

$$= \frac{4 + 2(2i) + i^2}{4 - i^2}$$

$$= \frac{4 + 4i - 1}{4 - (-1)}$$

$$= \frac{3 + 4i}{5}$$

19.

x	$y = x^2 - 4$	y
-3	$y = (-3)^2 - 4$	5
-2	$y = (-2)^2 - 4$	0
-1	$y = (-1)^2 - 4$	-3
0	$y = (0)^2 - 4$	-4
1	$y = (1)^2 - 4$	-3
2	$y = (2)^2 - 4$	0

See the graph on page A-38 in the textbook.

20.

x	$y = (x - 4)^2$	y
2	$y = (2 - 4)^2$	4
3	$y = (3 - 4)^2$	1
4	$y = (4 - 4)^2$	0
5	$y = (5 - 4)^2$	1
6	$y = (6 - 4)^2$	4

See the graph on page A-38 in the textbook.

Chapter 8 Test continued

21.

x	$y = (x + 3)^2 - 4$	y
-6	$y = (-6 + 3)^2 - 4$	5
-5	$y = (-5 + 3)^2 - 4$	0
-3	$y = (-3 + 3)^2 - 4$	-4
-1	$y = (-1 + 3)^2 - 4$	0
0	$y = (0 + 3)^2 - 4$	5

See the graph on page A-38 in the textbook.

22.

x	$y = x^2 - 6x + 11$	y
1	$y = 1^2 - 6(1) + 11$	6
3	$y = 3^2 - 6(3) + 11$	2
5	$y = 5^2 - 6(5) + 11$	6

See the graph on page A-38 in the textbook.

Chapter 9

Section 9.1

1. $x < -1$ or $x > 5$

    First we graph each inequality separately:

    $x < -1$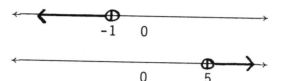

    $x > 5$

    Then, since the original two inequalities are connected by the word **or** , we graph everything that is on either graph above:

    $x < -1$ or $x > 5$

5. $x \le 6$ and $x > -1$

    First we graph each inequality separately:

    $x \le 6$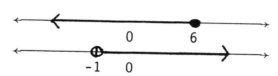

    $x > -1$

    Then, since the original two inequalities are connected by the word **and** , we graph all points that are common to the two graphs above:

    $x \le 6$ and $x > -1$

9. The graph on page A-38 of the text shows all points that are both greater than or equal to -2 and also less than or equal to 4.

13. The graph on page A-38 of the text shows all the points that are between -1 and 3 on the number line. That means the graph is all points that are both greater than -1 and also less than 3.

17. $3x - 1 < 5$ or $5x - 5 > 10$
    $3x < 6$ or $5x > 15$
    $x < 2$ or $x > 3$

    The graph of the solution is on page A-38 in your textbook.

21. $11x < 22$ or $12x > 36$
    $x < 2$ or $x > 3$

    The graph of the solution is on page A-38 in your textbook.

25. $2x - 3 < 8$ and $3x + 1 > -10$
$\qquad 2x < 11$ and $\qquad x > -11$

$\qquad x < \dfrac{11}{2}$ and $\qquad x > -\dfrac{11}{3}$

The graph of the solution is on page A-39 in your textbook.

29. $-1 \le x - 5 \le 2$
$\qquad 4 \le \quad x \quad \le 7 \qquad$ Add 5 to each member

The graph is on page A-39 in your textbook.

33. $-3 < 2x + 1 < 5$
$\qquad -4 < \quad 2x \quad < 4 \qquad$ Add -1 to each member

$\qquad -2 < \quad x \quad < 2 \qquad$ Multiply through by $\dfrac{1}{2}$

The graph is on page A-39 in your textbook.

37. $-7 < 2x + 3 < 11$
$\qquad -10 < \quad 2x \quad < 8 \qquad$ Add -3 to each member

$\qquad -5 < \quad x \quad < 4 \qquad$ Multiply through by $\dfrac{1}{2}$

41. $10 < x + 5 < 20$
$\qquad 5 < \quad x \quad < 15 \qquad$ Add -5 to each member

45. Let width = x, length = x + 4. Remember P = 2L + 2W

$\qquad 20 < \qquad\quad P \qquad\quad < 30$
$\qquad 20 < \qquad 2L + 2W \qquad < 30$
$\qquad 20 < 2(x + 4) + 2x < 30$
$\qquad 20 < \quad 2x + 8 + 2x \quad < 30 \qquad$ Distributive property
$\qquad 20 < \qquad 4x + 8 \qquad < 30 \qquad$ Simplify
$\qquad 12 < \qquad\quad 4x \qquad\quad < 22 \qquad$ Add -8 to each member

$\qquad 3 < \qquad\quad x \qquad\quad < \dfrac{11}{2} \qquad$ Multiply through by $\dfrac{1}{4}$

The width is between 3 inches and $\dfrac{11}{2}$ inches.

49. $-3 - 4(-2) = -3 + 8 \qquad$ Multiply
$\qquad\qquad\qquad = 5$

53. $5 - 2[-3(5 - 7) - 8] = 5 - 2[-3(-2) + -8] \qquad$ Add
$\qquad\qquad\qquad\qquad\quad = 5 - 2[6 + -8] \qquad$ Multiply
$\qquad\qquad\qquad\qquad\quad = 5 - 2[-2] \qquad$ Add
$\qquad\qquad\qquad\qquad\quad = 5 + 4 \qquad$ Multiply
$\qquad\qquad\qquad\qquad\quad = 9$

Section 9.1 continued

57. $\frac{1}{2}(4x - 6) = \frac{1}{2}(4x) - \frac{1}{2}(6)$ Distributive property

$\qquad\qquad = 2x - 3$ Multiply

61. $630 = 63 \cdot 10$

$\qquad = (7 \cdot 9) \cdot (2 \cdot 5)$

$\qquad = 7 \cdot 3 \cdot 3 \cdot 2 \cdot 5$

$\qquad = 2 \cdot 3^2 \cdot 5 \cdot 7$

Section 9.2

1. Step 1: Find the intercepts for the boundary.

x-intercept

when $\qquad\qquad y = 0$

$\qquad 2x - 3(0) = 6$

$\qquad\qquad 2x = 6$

$\qquad\qquad x = 3$

y-intercept

when $\qquad\qquad x = 0$

$\qquad 2(0) - 3y = 6$

$\qquad\qquad -3y = 6$

$\qquad\qquad y = -2$

Step 2: Since the inequality symbol is <, we graph the boundary as a broken line with x-intercept 3 and y-intercept -2.

Step 3: Using (0,0) in $2x - 3y < 6$, we have

$\qquad 0 - 0 < 6$

$\qquad 0 < 6$ A true statement

The solution set lies above the boundary.
The graph is on page A-39 in your textbook.

5. Step 1: Find the intercepts for the boundary.

x-intercept

when $\qquad\qquad y = 0$

$\qquad x - 0 = 2$

$\qquad\qquad x = 2$

y-intercept

when $\qquad\qquad x = 0$

$\qquad 0 - y = 2$

$\qquad\qquad -y = 2$

$\qquad\qquad y = -2$

Step 2: Since the original inequality symbol is $\leq$ we graph a solid line with x-intercept 2 and y-intercept -2.

Step 3: Using (0,0) in $x - y \leq 2$, we have

$\qquad 0 - 0 \leq 2$

$\qquad 0 \leq 2$ A true statement

The solution set lies above the boundary.
The graph is on page A-39 in your textbook.

9. Step 1: Find the intercepts for the boundary.

    x-intercept
    when $y = 0$
          $5x - 0 = 5$
             $5x = 5$
              $x = 1$

    y-intercept
    when $x = 0$
          $5(0) - y = 5$
               $-y = 5$
                $y = -5$

    Step 2: Since the inequality symbol is $\leq$ we have a solid line with
    x-intercept 1 and y-intercept -5 for our boundary.

    Step 3: Using (0,0) in $5x - y \leq 5$, we have

    $0 - 0 \leq 5$

    $0 \leq 5$ A true statement

    The solution set lies above the boundary.
    The graph is on page A-39 in your textbook.

13. Step 1: $x = 1$
    The boundary is $x = 1$, which is a vertical line.

    Step 2: The symbol is $>$, therefore the boundary is a solid line.

    Step 3: Using (0,0) in $x \geq 1$ we have

    $0 \geq 1$ A false statement

    The solution set lies to the right of the boundary.
    The graph is on page A-40 in your textbook.

17. Step 1: $y = 2$
    The boundary is $y = 2$, which is a horizontal line.

    Step 2: The symbol is $<$, therefore we have a broken line for the
    boundary.

    Step 3: Using (0,0) in $y < 2$, we have

    $0 < 2$ A true statement

    The solution set lies below the boundary.
    The graph is on page A-40 in your textbook.

21. Step 1: The boundary line is $y = 3x - 1$. The slope is 3 and the
    y-intercept is -1.

    Step 2: The symbol is $\leq$, therefore we have a solid line for the
    boundary.

    Step 3: Using (0,0) in $y \leq 3x - 1$, we have

    $0 \leq 3(0) - 1$
    $0 \leq -1$ A false statement

    The solution set lies above the boundary.
    The graph is on page A-40 in your textbook.

Section 9.2 continued

25. Step 1: The boundary is $y = -\frac{1}{2}x + 2$, which is a line with slope

$-\frac{1}{2}$ and y-intercept 2.

Step 2: The symbol is $\leq$, so the boundary is a solid line.

Step 3: Using $(0,0)$ in $y \leq -\frac{1}{2}x + 2$, we have,

$0 \leq -\frac{1}{2}(0) + 2$

$0 \leq 2$ A true statement

The solution set lies below the boundary.
The graph is on page A-40 in your textbook.

29. $7 - 3(2x - 4) - 8 = 7 - 6x + 12 - 8$ Distributive property
$= -6x + 11$

33. $8 - 2(x + 7) = 2$
$8 - 2x - 14 = 2$ Distributive property
$-2x - 6 = 2$ Simplify
$-2x = 8$ Add 6 to both sides
$x = -4$ Multiply each side by $-\frac{1}{2}$

37. $.15 = .1x > .25$
$-.1x > .10$ Add $-.15$ to both sides
$x < -1$ Multiply each side by $-.1$ and reverse the direction
of the inequality symbol.

See the graph on page A-40 in the textbook.

41.   Width = x
    Length = $3x + 5$. Remember $P = 2L + 2W$
Perimeter = 26 inches

$P = 2L + 2W$
$26 = 2(3x + 5) + 2x$
$26 = 6x + 10 + 2x$ Distributive property
$26 = 8x + 10$ Simplify
$16 = 8x$ Add -10 to both sides
$2 = x$ Multiply by $\frac{1}{8}$

Width is 2 inches, length is 11 inches.

Section 9.3

1. Let (3,7) be (x ,y) and (6,3) be (x ,y).

$$d = \sqrt{(6 - 3)^2 + (3 - 7)^2}$$
$$= \sqrt{(3)^2 + (-4)^2}$$
$$= \sqrt{9 + 16}$$
$$= \sqrt{25}$$
$$= 5$$

5. Let (3,-5) be (x , y) and (-2,1) be (x ,y).

$$d = \sqrt{(-2 - 3)^2 + [1 - (-5)]^2}$$
$$= \sqrt{(-5)^2 + (6)^2}$$
$$= \sqrt{25 + 36}$$
$$= \sqrt{61}$$

9.
$$\sqrt{13} = \sqrt{(1 - x)^2 + (5 - 2)^2}$$
$$13 = (1 - x)^2 + (3)^2$$
$$13 = 1 - 2x + x^2 + 9$$
$$13 = x^2 - 2x + 10$$
$$0 = x^2 - 2x - 3$$
$$0 = (x - 3)(x + 1)$$

x = 3 or   x = -1

13. Let (a,b) = (2,3) and r = 4, then the equation of a circle becomes:

$$(x - 2)^2 + (y - 3)^2 = 4^2$$
$$(x - 2)^2 + (y - 3)^2 = 16$$

17. Let (a,b) = (-5,-1) and r = $\sqrt{5}$, then the equation of a circle becomes:

$$[x - (-5)]^2 + [y - (-1)]^2 = (\sqrt{5})^2$$
$$(x + 5)^2 + (y + 1)^2 = 5$$

21. Let (a,b) = (0,0) and r = 2, then the equation of a circle becomes:

$$(x - 0)^2 + (y - 0)^2 = 2^2$$
$$x^2 + y^2 = 4$$

25. $(x - 1)^2 + (y - 3)^2 = 25$

The center is at (1,3) and the radius is 5.
The graph is on page A-41 in your textbook.

29.
$$(x + 1)^2 + (y + 1)^2 = 1$$
$$[x - (-1)]^2 + [y - (-1)]^2 = 1^2$$

The center is at (-1,-1) and the radius is 1.
The graph is on page A-41 in your textbook.

Section 9.3 continued

33. The line is vertical.
The graph is on page A-41 in your textbook.

37. Given m = -3, remember y = mx + b, therefore

$$y = -3x + b$$

Substituting x = -1, y = 4,

$$4 = -3(-1) + b$$
$$4 = 3 + b$$
$$1 = b$$

Substituting m = -3, b = 1 we have

$$y = -3x + 1$$

41. Given x + y = 20

Substituting y = 5x + 2 we have

$$x + (5x + 2) = 20$$
$$6x + 2 = 20$$
$$6x = 18$$
$$x = 3$$

Substituting x = 3 into y = 5x + 2, we have

$$y = 5(3) + 2$$
$$y = 17$$

The ordered pair is (3,17).

Section 9.4

1. $4^{1/2} = \sqrt{4} = 2$

5. $27^{1/3} = \sqrt[3]{27} = 3$

9. $18^{1/4} = \sqrt[4]{81} = 3$

13. $8^{2/3} = (8^{1/3})^2$ Separate exponents

    $= (\sqrt[3]{8})^2$ Write as cube root

    $= 2^2$ $\sqrt[3]{8} = 2$

    $= 4$ $2^2 = 4$

17. $16^{3/4} = (16^{1/4})^3$ Separate exponents

    $= (\sqrt[4]{16})^3$ Write as a fourth root

    $= 2^3$ $\sqrt[4]{16} = 2$

    $= 8$ $2^3 = 8$

21. $4^{2/3} = (4^{1/2})^3$ Separate exponents

$= (\sqrt{4}\,)^3$ Write as a square root

$= 2^3$ $\sqrt{4} = 2$

$= 8$ $2^3 = 8$

25. $(-32)^{1/5} = \sqrt[5]{-32}$ Write as a fifth root

$= -2$ $(-2)^5 = 32$

29. $16^{3/4} + 27^{2/3} = (16^{1/4})^3 + (27^{1/3})^2$ Separate exponents

$= (\sqrt[4]{16}\,)^3 + (\sqrt[3]{27}\,)^2$ Write as roots

$= 2^3 + 3^2$ $2^4 = 16,\ 3^3 = 27$

$= 8 + 9$ $2^3 = 8,\ 3^2 = 9$

$= 17$

33. $x^{1/4} \cdot x^{3/4} = x^{1/4 + 3/4}$ $(a^r a^s = a^{r+s})$

$= x^1$

$= x$

37. $\dfrac{a^{3/5}}{a^{1/5}} = a^{3/5 - 1/5}$ $\left(\dfrac{a^r}{a^s} = a^{r-s}\right)$

$= a^{2/5}$

41. $(9a^4 b^2)^{1/2} = 9^{1/2} \cdot a^{4(1/2)} \cdot b^{2(1/2)}$ $(a^r)^s = a^{rs}$

$= \sqrt{9} \cdot a^2 \cdot b$

$= 3a^2 b$

45. $25^{-1/2} = \dfrac{1}{25^{1/2}}$ $\left(a^{-r} = \dfrac{1}{a^r}\right)$

$= \dfrac{1}{\sqrt{25}}$

$= \dfrac{1}{5}$

49. $27^{-2/3} = \dfrac{1}{27^{2/3}}$ $\left(a^{-r} = \dfrac{1}{a^r}\right)$

$= \dfrac{1}{(27^{1/3})^2}$

$= \dfrac{1}{(\sqrt[3]{27}\,)^2}$

$= \dfrac{1}{3^2}$ $3^3 = 27$

$= \dfrac{1}{9}$ $3^2 = 9$

53. $\dfrac{x^4}{x^{-3}} = x^{4-(-3)}$

$\qquad = x^{4+3}$
$\qquad = x^7$

57. $20ab^2 - 16ab^2 + 6ab^2 = 10ab^2$

61. $2x^2(3x^2 + 3x - 1)$
$\qquad = 2x^2(3x^2) + 2x^2(3x) + 2x^2(-1)$
$\qquad = 6x^4 + 6x^2 - 2x^2$

65. $(2a^2 + 7)(2a^2 - 7)$
$\qquad = 4a^4 - 49 \qquad$ Difference of two perfect squares

Section 9.5

1. $(x + 4)(x - 4) = x^2 - 16$

5.
$$
\begin{array}{r}
x^2 - 2x + 4 \\
x + 2 \\
\hline
2x^2 - 4x + 8 \\
x^3 - 2x^2 + 4x \\
\hline
x^3 + 8
\end{array}
$$

9.
$$
\begin{array}{r}
x^2 - 4x + 16 \\
x + 4 \\
\hline
4x^2 - 16x + 64 \\
x^3 - 4x^2 + 16x \\
\hline
x^3 + 64
\end{array}
$$

13.
$$
\begin{array}{r}
x^2 - xy + y^2 \\
x + y \\
\hline
x^2y - xy^2 + y^3 \\
x^3 - x^2y + xy^2 \\
\hline
x^3 + y^3
\end{array}
$$

17. $x^3 + 7^3 = (x + 7)(x^2 - 7x + 49)$

21. $x^3 + 8 = x^3 + 2^3$
$\qquad = (x + 2)(x^2 - 2x + 4)$

25. $x^3 + 1 = x^3 + 1^3$
$\qquad = (x + 1)(x^2 - x + 1)$

29. $27x^3 - 8y^3 = (3x)^3 - (2y)^3$
$\qquad = (3x - 2y)(9x^2 + 6xy + 4y^2)$

Section 9.5 continued

33. $x^3 + 64y^3 = x^3 + (4y)^3$
$$= (x + 4y)(x^2 - 4xy + 16y^2)$$

37. $x^2 - 4x - 12$

Possible factors	Middle term when multiplied
$(x - 3)(x + 4)$	x
$(x + 3)(x - 4)$	$-x$
$(x + 2)(x - 6)$	$-4x$
$(x - 2)(x + 6)$	$4x$
$(x - 1)(x + 12)$	$11x$
$(x + 1)(x - 12)$	$-11x$

$x^2 - 4x - 12 = (x + 2)(x - 6)$

41. $5x^3 - 25x^2 - 30x$
$5x(x^2 - 5x - 6)$ Factor completely

Possible factors	Middle term when multiplied
$5x(x + 1)(x - 6)$	$-5x$
$5x(x - 1)(x + 6)$	$5x$
$5x(x + 2)(x - 3)$	$-x$
$5x(x - 2)(x + 3)$	x

$5x^3 - 25x^2 - 30x = 5x(x + 1)(x - 6)$

45.
$$x(x + 2) = 80$$
$$x^2 + 2x = 80 \qquad \text{Distributive property}$$
$$x^2 + 2x - 80 = 0 \qquad \text{Standard form}$$
$$(x + 10)(x - 8) = 0$$

$x + 10 = 0 \quad$ or $\quad x - 8 = 0$
$\qquad x = -10 \qquad\qquad x = 8$

Our two solutions are -10 and 8.

Section 9.6

1. When $y = 10$ and $x = 5$, $y = kx$ becomes

$$10 = k(5)$$
$$2 = k$$

Therefore, $y = 2x$. Since $x = 4$, we have

$$y = 2(4)$$
$$y = 8$$

5. When $y = -24$ and $x = 4$, $y = kx$ becomes

$$-24 = k(4)$$
$$-6 = k$$

Therefore, $y = -6x$. Since $y = -30$, we have

$$-30 = -6x$$
$$5 = x$$

256

9. When $y = 75$ and $x = 5$, $y = kx^2$ becomes
$$75 = k(5)^2$$
$$75 = 25k$$
$$3 = k$$

Therefore, $y = 3x^2$. Since $x = 1$, we have
$$y = 3(1)^2$$
$$y = 3$$

13. When $y = 5$ and $x = 2$, $y = \frac{k}{x}$ becomes
$$5 = \frac{k}{2}$$
$$10 = k$$

Therefore, $y = \frac{10}{x}$. Since $x = 5$, we have
$$y = \frac{10}{5}$$
$$y = 2$$

17. When $y = 5$ and $x = 3$, $y = \frac{k}{x}$ becomes
$$5 = \frac{k}{3}$$
$$15 = k$$

Therefore, $y = \frac{15}{x}$. Since $y = 15$, we have
$$15 = \frac{15}{x}$$
$$15x = 15$$
$$x = 1$$

21. When $y = 4$ and $x = 5$, $y = \dfrac{k}{x^2}$ becomes

$$4 = \frac{k}{5^2}$$

$$4 = \frac{k}{25}$$

$$100 = k$$

Therefore, $y = \dfrac{100}{x^2}$. Since $x = 2$, we have

$$y = \frac{100}{2^2}$$

$$y = \frac{100}{4}$$

$$y = 25$$

25. When $t = 42$ lbs. and $d = 2$ inches, $t = kd$ becomes

$$42 = k(2)$$
$$21 = k$$

Therefore $t = 21d$. Since $d = 4$, we have

$$t = 21(4)$$
$$t = 84$$

The tension is 84 lbs.

29. When $h = 20$ hours and $M = \$157$, $M = kh$ becomes

$$157 = k(20)$$
$$7.85 = k$$

Therefore, $M = 7.85h$. Since $h = 30$ hours, we have

$$M = 7.85(30)$$
$$M = \$235.50$$

He makes $235.50.

33. When $I = 30$ amps and $R = 2$ ohms, $I = \dfrac{k}{R}$ becomes

$$30 = \frac{k}{2}$$

$$60 = k$$

Therefore, $I = \dfrac{60}{R}$. Since $R = 5$ ohms, we have

$$I = \frac{60}{5}$$

$$I = 12$$

The current is 12 amps.

37. $\dfrac{x^2 - 25}{x + 4} \cdot \dfrac{2x + 8}{x^2 - 9x + 20} = \dfrac{(x + 5)(x - 5)}{(x + 4)(x - 5)(x - 4)} \, \dfrac{2(x + 4)}{}$

$$= \dfrac{2(x + 5)}{x - 4}$$

41. $\dfrac{1 - \dfrac{25}{x^2}}{1 - \dfrac{8}{x} + \dfrac{15}{x^2}} = \dfrac{x^2\left(1 - \dfrac{25}{x^2}\right)}{x^2\left(1 - \dfrac{8}{x} + \dfrac{15}{x^2}\right)}$

$$= \dfrac{x^2 - 25}{x^2 - 8x + 15}$$

$$= \dfrac{(x + 5)(x - 5)}{(x - 5)(x - 3)}$$

$$= \dfrac{x + 5}{x - 3}$$

45. Let x = amount of time to empty the pool with both pipes open.

$$\dfrac{1}{8} \quad - \quad \dfrac{1}{12} \quad = \quad \dfrac{1}{x}$$

Amount of water let in by inlet pipe	Amount of water let out by outlet pipe	Total amount of water in pool

$$24x\left(\dfrac{1}{8}\right) \quad - \quad 24x\left(\dfrac{1}{12}\right) \quad = \quad 24x\left(\dfrac{1}{x}\right) \qquad \text{Find LCD } 24x$$

$$3x \quad - \quad 2x \quad = \quad 24$$
$$x \quad = \quad 24 \text{ hours}$$

Remember to subtract inlet pipe from the outlet pipe because the pool is full and you are trying to empty the pool.

Section 9.7

1. Domain = {1, 3, 5}, Range = {2, 4, 6}

5. Domain = {1, 2, 3, 4}, Range = {1, 2, 3, 4}

9. Domain = {0, 2, 4}, Range = {2, $\dfrac{2}{3}$, $-\dfrac{2}{3}$}

13. $y = \dfrac{1}{x - 2}$

x = 2 makes the denominator zero:

Domain {x | x is real, x ≠ 2}

17. $y = \dfrac{4}{(x + 1)(x + 2)}$

  $x = -1$ and $x = -2$ makes the deonomiator zero:

  Domain $= \{x \mid x$ is real, $x \neq -1, x \neq -2\}$

21. $y = \dfrac{3}{x^2 - 9} = \dfrac{3}{(x + 3)(x - 3)}$

  $x = 3$ and $x = -3$ makes the denominator zero:

  Domain $= \{x \mid x$ is real, $x \neq 3, x \neq -3\}$

25. $y = \dfrac{x}{2x^2 - 5x - 3} = \dfrac{x}{(2x + 1)(x - 3)}$

  $x = -\dfrac{1}{2}$ and $x = 3$ makes the denominator zero:

  Domain $= \{x \mid x$ is real, $x \neq -\dfrac{1}{2}, x \neq 3\}$

29. $f(x) = x - 5$
  $f(2) = 2 - 5 = -3$
  $f(0) = 0 - 5 = -5$
  $f(-1) = -1 - 5 = -6$
  $f(a) = a - 5$

33. $f(x) = x^2 + 1$
  $f(2) = 2^2 + 1 = 5$
  $f(3) = 3^2 + 1 = 10$
  $f(4) = 4^2 + 1 = 17$
  $f(a) = a^2 + 1$

37. $f(x) = x^2 + 2x$
  $f(3) = 3^2 + 2(3) = 9 + 6 = 15$
  $f(-2) = (-2)^2 + 2(-2) = 4 - 4 = 0$
  $f(a) = a^2 + 2a$
  $f(a - 2) = (a - 2)^2 + 2(a - 2)$
  $\qquad\quad = a^2 - 4a + 4 + 2a - 4$
  $\qquad\quad = a^2 - 2a$

41. $\sqrt{49} = 7$

45. $\sqrt{\dfrac{2}{5}} = \dfrac{\sqrt{2}}{\sqrt{5}} \cdot \dfrac{\sqrt{5}}{\sqrt{5}}$

  $\qquad = \dfrac{\sqrt{10}}{5}$

49. $(\sqrt{6} + 2)(\sqrt{6} - 5) = (\sqrt{6})^2 + \sqrt{6}(-5) + 2\sqrt{6} + 2(-5)$

$$\qquad\qquad\qquad\quad\; \text{F} \qquad\quad \text{O} \qquad\quad \text{I} \qquad \text{L}$$

$$= 6 - 5\sqrt{6} + 2\sqrt{6} - 10$$

$$= -4 - 3\sqrt{6}$$

53. $\sqrt{2x - 5} = 3$

$\quad\; 2x - 5 = 9 \qquad$ Square both sides

$\qquad\quad 2x = 14$

$\qquad\qquad x = 7$

Check:

$$\sqrt{2(7) - 5} = 3$$

$$\sqrt{9} = 3$$

$$3 = 3$$

The solution is 7.

57. $\qquad\qquad 3x^2 = 4x + 2$

$3x^2 - 4x - 2 = 0 \qquad$ Standard form

When a = 3, b = -4 and c = -2,

$$x = \frac{-b \pm \sqrt{b^2 - 4ac}}{2a} , \qquad \text{the quadratic formula, becomes:}$$

$$= \frac{-(-4) \pm \sqrt{(-4)^2 - 4(3)(-2)}}{2(3)}$$

$$= \frac{4 \pm \sqrt{16 + 24}}{6}$$

$$= \frac{4 \pm \sqrt{40}}{6}$$

$$= \frac{4 \pm 2\sqrt{10}}{6}$$

$$= \frac{2(2 \pm \sqrt{10})}{6}$$

$$= \frac{2 \pm \sqrt{10}}{3}$$

The two solutions are $\dfrac{2 + \sqrt{10}}{3}$ and $\dfrac{2 - \sqrt{10}}{3}$

61. $(4 + 3i) - (2 - 5i) = 4 + 3i - 2 + 5i \qquad$ Distributive property

$$= (4 - 2) + (3i + 5i)$$

$$= 2 + 8i$$

Chapter 9 Review

1. See the graph on page A-43 in the textbook.

5. $-5x \geq 25$ or $2x - 3 \geq 9$

   $x \leq -5$ $\qquad$ $2x \geq 12$

   $\qquad\qquad\qquad$ $x \geq 6$

   See the graph on page A-43 in the textbook.

9. $-1 \leq 3x + 5 \leq 8$

   $-5 \qquad -5 \quad -5$

   $\dfrac{-6}{3} \leq \dfrac{3x}{3} \leq \dfrac{3}{3}$

   $-2 \leq x \leq 1$

   See the graph on page A-43 in the textbook.

13. $x \geq -3$ $\qquad$ See the graph on page A-43 in the textbook.

17. Letting $(x_1, y_1) = (2,6)$ and $(x_2, y_2) = (-1,5)$

   $$\begin{aligned} d &= \sqrt{(x_2 - x_1) + (y_2 - y_1)} \\ &= \sqrt{(-1 - 2)^2 + (5 - 6)^2} \\ &= \sqrt{(-3)^2 + (-1)^2} \\ &= \sqrt{9 + 1} \\ &= \sqrt{10} \end{aligned}$$

21. Letting $(x_1, y_1) = (-3,7)$ and $(x_2, y_2) = (-3,-2)$

   $$\begin{aligned} d &= \sqrt{(x_2 - x_1)^2 + (y_2 - y_1)^2} \\ &= \sqrt{[-3 - (-3)]^2 + (-2 - 7)^2} \\ &= \sqrt{0^2 + (-9)^2} \\ &= \sqrt{81} \\ &= 9 \end{aligned}$$

25. Letting $(a,b) = (3,1)$ and $r = 2$

   $(x - a)^2 + (y - b)^2 = r^2$

   $(x - 3)^2 + (y - 1)^2 = 4$

29. $x^2 + y^2 = 4$

   $(x + 0)^2 + (y + 0)^2 = 2^2$

   Center $(0,0)$, $r = 2$

33. $36^{1/2} = \sqrt{36} = 6$

37. $(-8)^{2/3} = [(-8)^{1/3}]^2$
$$= (\sqrt[3]{-8})^2$$
$$= (-2)^2$$
$$= 4$$

41. $(16y^{20})^{1/4} = 16^{1/4}(y^{20})^{1/4}$
$$= \sqrt[4]{16}\,(y^5)$$
$$= 2y^5$$

45. $a^3 - 7^3 = (a - 7)(a^2 + 7a + 49)$

49. $27x^3 + y^3 = (3x)^3 + y^3 = (3x + y)(9x^2 - 3xy + y^2)$

53. When $y = -20$ and $x = 4$, $y = kx$ becomes
$$-20 = k(4)$$
$$-5 = k$$

Therefore, $y = -5x$. Since $x = 7$, we have
$$y = -5(7)$$
$$y = -35$$

57. When $y = 8$ and $x = 36$, $y = \dfrac{k}{\sqrt{36}}$ becomes

$$8 = \frac{k}{\sqrt{36}}$$

$$8 = \frac{k}{6}$$

$$48 = k$$

Therefore, $y = \dfrac{48}{\sqrt{x}}$. Since $x = 144$, we have

$$y = \frac{48}{\sqrt{144}}$$

$$y = \frac{48}{12}$$

$$y = 4$$

61. $y = 3x - 1$ for $x = 2$ $y = 3x - 1$ for $x = 4$
       $y = 3(2) - 1$ $y = 3(4) - 1$
       $y = 5$ $y = 12 - 1$
                                     $y = 11$

       $y = 3x - 1$ for $x = 6$
       $y = 3(6) - 1$
       $y = 18 - 1$
       $y = 17$

       Domain = $\{2,4,6\}$ Range = $\{5,11,17\}$

65. $y = \dfrac{5}{x^2 = 8x + 15}$

       $x = 3$ and $x = 5$ makes the denominator zero:

       Domain = $\{x \mid x$ is real, $x \neq 3, x \neq 5\}$

69. $f(x) = 2x^2 + 1$
      $f(-2) = 2(-2)^2 + 1 = 2(4) + 1 = 8 + 1 = 9$
       $f(2) = 2(2)^2 + 1 = 2(4) + 1 = 8 + 1 = 9$
       $f(a) = 2a^2 + 1$

73. $f(x) = x^3 + 1$
      $f(-2) = (-2)^3 = -8 + 1 = -7$
       $f(2) = 2^3 + 1 = 8 + 1 = 9$
       $f(a) = a^3 + 1$

Chapter 9 Test

1. x < 2 or x > 3

    Graph each inequality separately

        x < 2

        x > 3

    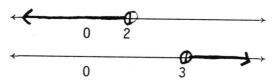

    The graph of the solution is on page A-44 in the textbook.

2. x > -2 and x < 3

    Graph each inequality separately.

        x > -2

        x < 3

    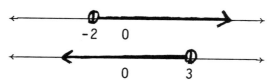

    The graph of the solution is on page A-44 in the textbook.

3. 3 - 4x $\geq$ -5 or 2x $\geq$ 10

    Graph each inequality separately.

        3 - 4x $\geq$ -5

          -4x $\geq$ -8

            x $\leq$ 2

    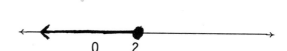

          2x $\geq$ 10

            x $\geq$ 5

    The graph of the solution is on page A-44 in the textbook.

4. -7 < 2x - 1 < 9
   -6 < 2x   < 10   Add 1 to each member
   -3 < x   < 5   Divide each by 2

   The graph of the solution is on page A-44 in the textbook.

Chapter 9 Test continued

5. $y < -x + 4$

Step 1: $y = -x + 4$
Graph the line with slope = -1 and y-intercept = 4.

Step 2: The symbol is <, therefore we have a broken line.

Step 3: Using (0,0) in $y < -x + 4$, we have

$0 < -0 + 4$
$0 < 4$ A true statement

The solution set lies below the boundary.

The graph is on page A-44 in the textbook.

6. $3x - 4y \geq 12$

Step 1: $3x - 4y = 12$

x-intercept
$y = 0$
$3x - 4(0) = 12$
$3x = 12$
$x = 4$

x-intercept at (4,0)

y-intercept
$x = 0$
$3(0) - 4y = 12$
$-4y = 12$
$y = -3$

y-intercept a (0,-3)

Step 2: The symbol is $\geq$, therefore we have a solid line.

Step 3: Using (0,0) in $3x - 4y \geq 12$, we have

$3(0) - 4(0) \geq 12$
$0 \geq 12$ A false statement

The solution set lies below the boundary.

The graph is on page A-44 in the textbook.

7. Let (1,2) be (x_1,y_1) and (4,6) be (x_2,y_2)

$d = \sqrt{(x_2 - x_1)^2 + (y_2 - y_1)^2}$
$= \sqrt{(4 - 1)^2 + (6 - 2)^2}$
$= \sqrt{3^2 + 4^2}$
$= \sqrt{9 + 16}$
$= \sqrt{25}$
$= 5$

8. Let $(-4,1)$ be (x_1,y_1) and $(2,5)$ be (x_2,y_2)

$$d = \sqrt{(x_2 - x_1)^2 + (y_2 - y_1)^2}$$
$$= \sqrt{(2 - (-4))^2 + (5 - 1)^2}$$
$$= \sqrt{(2 + 4)^2 + 4^2}$$
$$= \sqrt{6^2 + 4^2}$$
$$= \sqrt{36 + 16}$$
$$= \sqrt{52}$$
$$= \sqrt{4 \cdot 13}$$
$$= 2\sqrt{13}$$

9. Let $(a,b) = (2,-3)$ and $r = 5$

$$(x - a)^2 + (y - b)^2 = r^2$$
$$(x - 2)^2 + (y - (-3))^2 = 5^2$$
$$(x - 2)^2 + (y + 3)^2 = 25$$

10. $(x - 2)^2 + (y - 3)^2 = 3^2$

The center is at $(2,3)$ and the radius is 3.

The graph is on page A-44 in the textbook.

11. $25^{1/2} = \sqrt{25} = 5$

12. $8^{2/3} = (8^{1/3})^2$ Separate exponents

$\quad = (\sqrt[3]{8})^2$ Write as cube root

$\quad = 2^2 \qquad \sqrt[3]{8} = 2$

$\quad = 4 \qquad 2^2 = 4$

13. $x^{3/4}x^{1/4} = x^{3/4 + 1/4}$

$\qquad = x^1$

$\qquad = x$

14. $(27x^6y^9)^{1/3} = 27^{1/3}x^{6(1/3)}y^{9(1/3)}$

$\qquad = \sqrt[3]{27}\, x^2 y^3$

$\qquad = 3x^2y^3$

15. $x^3 - 7^3 = (x - 7)(x^2 + 7x + 49)$

16. $x^3 + 27 = x^3 + 3^3$

$\qquad = (x + 3)(x^2 - 3x + 9)$

17. $(2x)^3 + (5y)^3 = (2x + 5y)(4x^2 - 10xy + 25y^2)$

18. $8x^3 - 27y^3 = (2x)^3 - (3y)^3$
$ = (2x - 3y)(4x^2 + 6xy + 9y^2)$

19. $y = kx^2$ when $x = 3$ and $y = 36$
$36 = k3^3$
$36 = 9k$
$4 = k$ therefore

$y = 4x^2$ substitute $x = 5$
$y = 4(5)^2$
$y = 4(25)$
$y = 100$

20. $y = \dfrac{k}{x}$ when $x = 3$ and $y = 6$

$6 = \dfrac{k}{3}$

$18 = k$ therefore

$y = \dfrac{18}{x}$ substitute $x = 9$

$y = \dfrac{18}{9}$

$y = 2$

21. Domain = $\{1,3,5\}$
Range = $\{2,4,6\}$

22. $y = \dfrac{5}{x^2 - 25} = \dfrac{5}{(x + 5)(x - 5)}$

$x = 5$ and $x = -5$ makes the denominator zero:

Domain = $\{x \mid x$ is a real, $x \neq 5, x \neq -5\}$

23. $f(x) = 3x^2 - 1$
$f(3) = 3(3)^2 - 1$
$ = 3(9) - 1$
$ = 27 - 1$
$ = 26$

24. $f(x) = 3x^2 - 1$
$f(-2) = 3(-2)^2 - 1$
$ = 3(4) - 1$
$ = 12 - 1$
$ = 11$

Chapter 9 Test continued

25. $g(x) = 5x + 2$
     $g(0) = 5(0) + 2$
     $ = 2$

26. $g(x) = 5x + 2$
     $g(-5) = 5(-5) + 2$
     $ = -25 + 2$
     $ = -23$